Evolution Made to Order

Plant Breeding and Technological Innovation in Twentieth-Century America

HELEN ANNE CURRY

The University of Chicago Press
Chicago and London

Helen Anne Curry is the Peter Lipton Lecturer in History of Modern Science and Technology at the University of Cambridge.

The University of Chicago Press, Chicago 60637
The University of Chicago Press, Ltd., London
© 2016 by The University of Chicago
All rights reserved. Published 2016.
Printed in the United States of America

25 24 23 22 21 20 19 18 17 16   1 2 3 4 5

ISBN-13: 978-0-226-39008-6 (cloth)
ISBN-13: 978-0-226-39011-6 (e-book)
DOI: 10.7208/chicago/9780226390116.001.0001

Library of Congress Cataloging-in-Publication Data

Names: Curry, Helen Anne, author.
Title: Evolution made to order : plant breeding and technological innovation in twentieth-century
    America / Helen Anne Curry.
Description: Chicago ; London : University of Chicago Press, 2016. | ©2016 | Includes bibliographical
    references and index.
Identifiers: LCCN 2015047402 | ISBN 9780226390086 (cloth : alkaline paper) | ISBN 9780226390116
    (e-book)
Subjects: LCSH: Plant mutation breeding—United States—History—20th century. | Plant mutation
    breeding—Social aspects—United States. | Plant genetic engineering—Genetic engineering—United
    States—History—20th century. | Plant genetic engineering—Social aspects—United States.
Classification: LCC SB83.C87 2016 | DDC 631.5/233—dc23 LC record available at http://lccn.loc.gov/
    2015047402

♾ This paper meets the requirements of ANSI/NISO Z39.48-1992 (Permanence of Paper).

For my parents

Contents

Abbreviations

AAAS: American Association for the Advancement of Science

AEC: Atomic Energy Commission

AIP BROOKHAVEN: Brookhaven Director's Office Files (microfilm), Niels Bohr Library and Archives, American Institute of Physics, College Park, Maryland

BLAKESLEE PAPERS: Albert Francis Blakeslee Papers, American Philosophical Society, Philadelphia, Pennsylvania

BNL: Brookhaven National Laboratory

CERLO: Committee on Effects of Radiation upon Living Organisms

CERLO FILES: Committee on Effects of Radiation upon Living Organisms Files, Division of Biology and Agriculture Collection, Archives of the National Academy of Sciences, Washington, DC

CIW: Carnegie Institution of Washington

CIW FILES: Carnegie Institution of Washington Files, Cold Spring Harbor Archives, Cold Spring Harbor, New York

CNEN: Comitato Nazionale per l'Energia Nucleare

CORNELL DPB: Department of Plant Breeding Records, Division of Rare and Manuscript Collections, Cornell University, Ithaca, New York

CSHL: Cold Spring Harbor Laboratory

CU HORTICULTURE: Horticulture Department Records, Clemson University Archives, Special Collections, Clemson University, Clemson, South Carolina

FAO: Food and Agriculture Organization of the United Nations

GE: General Electric

GOODSPEED PAPERS: Thomas Harper Goodspeed Papers, Bancroft Library, University of California, Berkeley

GM: genetically modified

IAEA: International Atomic Energy Agency

MULLER PAPERS: Hermann Joseph Muller Papers, Lilly Library, Indiana University, Bloomington

NARA RG 54: Records of the Bureau of Plant Industry, Soils, and Agricultural Engineering, US National Archives and Records Administration II, College Park, Maryland

NARA RG 326-G: Records of the Atomic Energy Commission, Photographic Prints, AEC Installations, Facilities, Personalities and Activities, 1947–1972, US National Archives and Records Administration II, College Park, Maryland

NRC: National Research Council

OPENNET: US Department of Energy OpenNet Database, https://www.osti.gov/opennet

SEARS PAPERS: Ernest R. Sears Papers, Western Historical Manuscript Collection, Columbia, Missouri

SETCHELL PAPERS: William Setchell Papers, University and Jepson Herbaria Archives, University of California, Berkeley

SINGLETON PAPERS: W. Ralph Singleton Papers, Small Special Collections Library, University of Virginia, Charlottesville

SPARROW PAPERS: A. H. Sparrow Papers, Special Collections, University of Tennessee, Knoxville

STADLER PAPERS: Lewis J. Stadler Papers, Western Historical Manuscript Collection, Columbia, Missouri

TAYLOR PAPERS: Frank J. Taylor Papers, Department of Special Collections, Stanford University, Stanford, California

UM AGRONOMY: Administrative Records, College of Agriculture, Food, and Natural Resources & Department of Agronomy, Archives of the University of Missouri, Columbia

UN: United Nations

USDA: United States Department of Agriculture

UT-AEC: University of Tennessee–Atomic Energy Commission

UTAES: University of Tennessee Agricultural Experiment Station

UT NRC: Committee on Effects of Radiation upon Living Organisms, National Research Council Reports (photocopies), Special Collections, University of Tennessee, Knoxville

To create a new variety, a plant breeder first needs to find the right genes. In the nineteenth century, American breeders hoping for wheat that could survive harsh conditions on the northern plains looked to Europe to discover plants with characteristics that made them tolerant of drought and cold. Corn breeders of the 1940s, searching for a way to make the production of hybrids more economical, began transferring a trait for partial sterility found in one aberrant plant in a Texas cornfield into the vast majority of commercially grown corn in the United States. In the 1990s, breeders used molecular technologies to transfer genetic material from an arctic flounder to the supermarket tomato, anticipating that the fish gene would lead to a more frost-resistant and fridge-ready vegetable. In each of these cases, like so many others, breeders sought heritable variations wherever they existed in the world. In some instances, they used variant plants as the foundation of a new line, while in others they incorporated new genetic material into existing varieties via methods such as hybridization and genetic engineering. This basic approach has been extremely effective. Plant breeders have transformed agricultural production around the world, especially in the last one hundred years.[1]

Despite this apparent success, breeders have often expressed a desire to escape the constraints imposed by their reliance on variation as found in nature. Why should they have to wait for some unknown and unreliable cause to provoke spontaneous mutation in the bud of an apple, leading to a sweeter fruit? Why should they rely on a hope that some desired genetic variation already exists somewhere, in an extant wheat variety or an individual corn plant or even a flounder? And why should they go through the painstaking effort of transferring desired traits from one line or breed to another? Surely if one could find a way to produce variations on demand, in the varieties

USE X-RAY TO GET BIGGER CORN CROP

TRUNDLING an X-ray machine on wheels up and down the rows of an experimental garden, Dr. L. J. Stadler, of the Missouri College of Agriculture, is seeking to produce new species of farm crops by subjecting his plants to treatment by the invisible rays. Strange alterations take place in corn and other growing plants when they are bombarded by X-rays, laboratory tests have previously shown. Dr. Stadler expects improved varieties to result from this practical experiment. His portable apparatus receives electric current by a long cable.

Dr. L. J. Stadler, of Missouri College of Agriculture, and his portable X-ray machine used to get better farm crops

FIGURE 1. The geneticist Lewis Stadler conducts an experiment in a cornfield with a portable x-ray machine, reportedly in hopes of creating "better farm crops." From *Popular Science Monthly*, January 1932, 47.

in which they were absent, then all these time-consuming processes could be avoided. In the 1920s, American plant breeders felt particularly aggrieved about the lack of a method for producing heritable variation on demand—and particularly hopeful that the still young science of genetics would come up with just the right tool.

Enter the x-ray machine. Although x-ray devices had been around since the turn of the century, deployed by doctors and dentists, physicists and industrialists, they were newly celebrated in 1927—by breeders. A handful of scientists had demonstrated that the rays sometimes produced heritable variations in bacteria, flies, corn, tobacco, and still other organisms. These successes suggested that x-rays could be used to intervene in evolution itself, assuming, that is, that the heritable alterations, or mutations, were equivalent to changes that occurred spontaneously in nature. That in turn suggested their enormous potential in agricultural production (fig. 1). As one breathless account of the x-ray research explained, until the effect had been uncovered, "nobody dreamed that X-rays might revolutionize agriculture. Now we know that X-rays can accomplish in a few weeks what nature has been taking millions of years to do. They can speed up evolution until you get dizzy thinking about it."[2] In truth it was not yet clear whether x-ray radiation could be used in this way, but breeders and other experimenters certainly gave it their best shot. They tested the technique on all kinds of plants from rice varieties to gladiolus flowers to redwood trees.

Although the evolution revolution never materialized, x-ray machines were only the first of a series of strange tools celebrated by breeders in the middle decades of the twentieth century. Plant breeders also worked with chemical solutions and radioisotopes, similarly envisioning that these would enable them to reshape plants—and perhaps even animals—to specification. According to scientific and popular reports, scientists would use these tools to generate heritable variation "at will," which would in turn allow breeders to develop agricultural organisms "to order." No more exhaustive searches for natural variations. No more complex integration of desired variation into established breeds through hybridization and selection. Breeders would instead alter genes and chromosomes directly, transforming fruits, grains, vegetables, and flowers with unprecedented efficiency.

This book tells the story of these early technologies for altering plant genes and chromosomes. It explores the longstanding hopes and expectations that fostered pursuit of new genetic technologies, and charts the biological research in heredity and evolution that contributed to their eventual innovation. It traces the uses of these in agriculture and horticulture, to see what breeders hoped to gain by particular methods and to learn what they actually achieved. It delves into the reports, sober and sensational alike, that appeared in newspapers and magazines, to reveal the perceived significance of these genetic technologies outside the laboratory and beyond the farm field. They were clearly considered significant. In the middle decades of the twentieth century, the physical nature of genes was not yet fully described or understood, much less were the sophisticated tools needed for the manipulation of specific genes available. In this context, the ability to force mutations—however random—was one of the most promising applications of the developing science of genetics.

Although readers will encounter in these stories many familiar sites and subjects from the history of biology, one of the central arguments of this book is that it is impossible to understand early genetic technologies apart from the broader history of American technology and innovation. Genetic technologies were completely entangled with other areas of innovation, both in their material production and in the outcomes anticipated from them. In the early decades of the twentieth century, industrial research produced ever-more-powerful x-ray technologies, useful in medicine and manufacturing. It was just possible that an x-ray device precisely calibrated for plant breeding might be produced as well. In the 1930s and 1940s, chemical technologies cured diseases and created fertile farm fields—surely they could also produce genetically enhanced crops. After World War II, nuclear power promised a future of energy too cheap to meter; why not expect nuclear-aided plant breeding to feed hungry mouths worldwide, more quickly and cheaply than traditional approaches? In capturing this entanglement, I show that many Americans envisioned and enacted the process of innovating living organisms, in this case new breeds of agricultural crops and garden flowers, little differently from that of innovating any other modern industrial product.

The story told of American technological innovation in the first half of the twentieth century is most often one of celebration and anticipation. By 1900, the United States was no longer a rural, agricultural society but an industrial and, increasingly, urban one. Railroad lines and telegraph wires crisscrossed the country, ensuring a continuous flow of consumer goods, agricultural products, financial transactions, and personal and business communications from coast to coast. American ingenuity had already produced many marvels, things like electric lighting, telephones, and phonographs, and the pace of innovation continued unabated. From automobiles to radios to electric home appliances, Americans of the 1910s, 1920s, and 1930s encountered an ever-expanding array of consumer goods, all of which promised to make life easier and more enjoyable. Technological change sometimes generated fears, such as of workers made redundant by machine labor (confirmed for some by the Great Depression) or of landscapes destroyed in the relentless pursuit of speed and power; however, these were often overshadowed by utopian visions of a society made ever more efficient and productive through the application of scientific and technical knowledge. This was true even as Americans sought a way out of the economic depression of the 1930s, and later as they faced the specter of a second world war.[3]

Optimism about technology extended to American agricultural production, which had experienced a significant transformation in the nineteenth century with the opening up of the West to farming, the creation of national markets for agricultural products (in large part a result of rail and telegraph networks), and the introduction of laborsaving machinery. An extended period of depressed prices reversed in 1897, meaning that incomes for farmers at the turn of the century were on the rise. And the future looked rosier still. There was the promise of further technological improvement, via farm equipment such as harvesters and combines, the construction of dams and other irrigation technologies in the arid west, and the application of scientific principles to crop breeding and methods of cultivation. When agricultural incomes plunged after the Great War, sending American farmers into a spiraling depression while the rest of the country boomed, many looked to science and technology for the solutions to the crisis. They sought new crops to alleviate the problems of overproduction, novel industrial uses of farm wastes, and better ways of organizing agricultural markets. They purchased tractors in large numbers and adopted electrical devices like milking machines and brooders for raising chickens. In the ensuing decades, farmers continued to seek out and adopt technologies, such that the dominant mode of agricultural production by the postwar years—large-scale, intensive, and dependent on machines, synthetic chemicals, and petroleum—could only be accurately described as "industrial."[4]

Although thoroughgoing industrialization was in most cases a post-1945 phenomenon, even from the turn of the century, agricultural production in America comprised a complex technological system.[5] Land-grant colleges and agricultural experiment stations generated both knowledge about farming practices and technical specialists in subjects ranging from livestock management to agricultural chemistry to home economics. Farmers who raised crops and animals for market relied on these specialists and the knowledge they produced; they also depended increasingly on American industrial production and the range of manufactured equipment and tools it made available. Railroads and other transportation systems facilitated the shipment of goods, whether commercially produced seeds on their way to farmers or the vegetables grown from these seeds on their way to markets. Meanwhile print and electric communications systems connected farmers to one another and to the world. Interleaved and often coextensive with this agricultural system was an infrastructure dedicated to smaller-scale agricultural and horticultural production, including home gardening. Many commercial seedhouses catered to small-scale cultivators as well as to larger operations. The production, marketing, and sale of flower and vegetable seeds also relied on advances in breeding and plant propagation and on the growing rail and mail networks.[6] Within these interlocking systems, there existed many niches in which a novel innovation—an economic theory, a piece of farm machinery, a variety of some economic crop or garden flower—could flourish. This in turn created many opportunities for those who would generate such innovations to make their careers.

This book charts the activities of individuals who hoped to innovate tools aimed at one specific end: physical manipulation of the genes and chromosomes of plants. These would-be innovators included university professors of genetics, agronomists and breeders at state agricultural experiment stations, geneticists and plant biologists at independent research laboratories, breeders at commercial seedhouses, engineers at industrial research laboratories, and still others. The diversity seen among their institutional affiliations and occupations suggests that the development of methods for direct genetic manipulation was of interest for many different reasons. A university geneticist might find acclaim for having cracked a great mystery of biology. Ditto for an agronomist, who would then also be able to improve crucial economic crops. A commercial flower breeder could stay one step ahead of the competition in the cutthroat market for flower novelties. An industrial engineer might envision patenting and profiting from a machine that churned out genetic alterations on demand. Regardless of their motivation, all saw significant professional rewards to having produced or perfected a technology for manipulating heredity in plants. And most believed in the eventual transformative power—for agriculture, for the economy, for American lives—of such an innovation.

Historians have written a great deal about the persistent vision of achieving human control over the form and evolution of living organisms and the effects of this vision on biological research. Histories of eugenics in particular, with their elaboration of the many and misguided efforts made to control human heredity, have indicated just how deep and enduring this vision was, and is, and the extremes to which faith in technical knowledge has carried efforts to intervene in heredity and evolution.[7] Eugenics is of course not the only story. Many areas of life sciences research and even technological development in the twentieth century and earlier were premised on the idea that living organisms, from mold strains to wheat varieties to human beings, are plastic or can be made plastic to meet scientific, social, or commercial aims. Historians writing on subjects as diverse as beer brewing, agricultural breeding, landscaping and gardening, experimental physiology, tissue culture, and genetics and molecular biology have offered accounts of these as efforts to technologize biological systems. These histories have shown us that experimental biology has always incorporated a vision of technological control, that attempts at engineering and synthesis have long been legitimate modes of biological research, and that living organisms at many points over the past century have been incorporated into—or even themselves been transformed into—technologies.[8]

The use of radiation and chemicals in genetics and plant breeding shares in this history.[9] These, too, were seen as means of controlling evolution and heredity. Above all, they seemed to offer a way to accelerate the pace of evolutionary change, by placing the appearance of heritable variation under human control. "It brings something into the field of plant breeding that we have always wanted but never had," the geneticist Albert Blakeslee enthused in 1938 of a chemical technique for multiplying the number of chromosomes in many kinds of plants. "We can get the chromosome to double up almost at will and thus are able to speed up and to control the processes of evolution."[10] Even more compelling was the fact that this control was based in direct physical or chemical manipulation of genetic material. According to one account, x-ray breeding worked because "the X-rays act like a carpenter with a saw and hammer entering the strings of chromosomes that give living things their colors, shapes and flavors. The rays cut some of these chromosomes out of their natural positions and fix them elsewhere. The result is a changed plant or animal."[11] In the early decades of the twentieth century this ability to "enter the strings of chromosomes" was a startling achievement, one that seemed to set these tools apart from methods that had preceded them.

As this suggests, *Evolution Made to Order* is a book about the biological sciences, especially genetics, cytology, and breeding, and about a shared vision of biological control. But it is also a book about technology. More specifically, it is a history of innovation.[12] The chapters that follow chart the efforts made by many individuals to produce novel technologies for use in genetics research and plant breeding. And because plant breeding is itself an innovative activity, one that results in new varieties of agricultural crops, fruit trees, and garden flowers, this book is in fact a nested history of innovation. It is a history of innovations (in breeding techniques and technologies) whose primary purpose was to generate further innovations (in crops, trees, and flowers).

One way to tell a history of innovation in plant breeding would be to focus on technologies deemed historically important and the individuals credited with bringing them into being. For example, one might recount the work of famous breeders, like the turn-of-the-century horticulturist Luther Burbank or the celebrated Green Revolution wheat breeder Norman Borlaug, and the processes by which they created marvelous new crops.[13] Or one could focus on the innovations in plant breeding that produced lasting changes in breeders' methods, like the double-cross hybrid technique in corn or transgenic engineering.[14] Yet one of the most important critiques of the history of technology made in the past thirty-odd years (and, in some cases, continuing into the present) was that the field was too centered on specific innovations and on the work of professional innovators. Taking a narrow view of who and what is important in the creation of a "successful" technology misses many crucial elements, such as the history of technologies in use, the role of diverse users in shaping technological development, and the importance of the larger structures and systems in which individual innovations are embedded.[15] Even the focus on technologies that succeed skews our picture of technological development. Investigations of so-called failed innovations, those that did not persist or produce despite high expectations, offer insight into the techno-logical aspirations of engineers and corporations and the role of consumer expectations in shaping technological outcomes.[16]

What follows, therefore, is not an account of well-known plant breeders and novel plant varieties that proved wildly successful. Nor is it a history of technologies that revolutionized plant breeding. It is a history of three inno-vations in plant breeding methods that proved far less useful in generating new plants than was initially hoped. In charting this history, I put aside the assumption that these were simply failures and therefore of little use in un-derstanding twentieth-century innovations in plant breeding. It is true that they never lived up to the outsized expectations many had for them. X-ray machines did not shorten the time required to breed a new strain of corn. Chemical manipulation of chromosomes did not lead to never-before-seen fruit hybrids. Exposing soybeans to gamma rays from an isotope of cobalt-60 did not immediately generate disease-resistant varieties. Exploring the efforts made to realize these outcomes nonetheless reveals a great deal, not only about the history of plant breeding and genetics, but also about long-held, widely shared expectations for genetic technologies and about the intersection of plant breeding with other technological domains.

Furthermore, even though these technologies never produced revolution-ary effects and, in most cases, barely made a difference in dominant meth-ods of breeding, that does not necessarily mean that they "failed." All saw use. All led to the production of new plant varieties. All satisfied, at one time or another, the varied needs of various users—needs that were not always principally about the creation of better agricultural crops. Industrial patrons advocated the use of technologies in which they had a vested economic in-terest (for example, x-rays), while the US government encouraged the use of those in which it had a significant political investment (atomic energy). Professional geneticists and breeders benefited when technologies were diffi-cult to access or demanded technical skills (double-cross hybrids), and they championed these approaches. Amateurs, on the other hand, extended the use of more accessible and easy-to-use tools (chemical treatment). Even pro-fessional breeders differed from one another in their preferences for different technologies, depending on the kinds of plants they worked with. The relative ease of introducing to the market a novel flower, such as a rose or a marigold, compared with an agricultural commodity like corn or wheat, increased the willingness of commercial flower breeders to try out, and find success with, almost any new breeding technology. In short, the expectations of different users, and the different demands they faced within particular institutional settings, influenced the attention they gave to each innovation, the ways in which they put it to use, and the likelihood of its persisting in spite of less-than-stellar outcomes.

Today, the term "biotechnology" is most often used to refer to tools, products, and institutions that have emerged since the 1970s when methods first became available for splicing and transferring genetic material from the genome of one organism to that of another. Biotechnology in this sense encompasses the transgenic mice created for medical research, so-called Roundup Ready soybeans and other crop varieties genetically engineered to be herbicide resistant, and bacteria modified to produce human proteins for therapeutic purposes—along with the techniques and organizations behind the creation of these and many similar organisms. But biotechnology can also refer to the use of living organisms in any industrial product or process, regardless of whether it involves transgenic or other molecular genetic manipulation. Examples might include the use of yeast in brewing beer or making soy sauce, the cultivation of improved crop plants and the domestication of animals, the culturing of mold to produce penicillin, the propagation of tissue cultures to manufacture vaccines, and so on. These, too, are biological technologies in which living organisms or materials are transformed into tools for achieving human ambitions or meeting human needs. Biotechnology encompasses post-1970 molecular genetic technologies, then, but it is not limited to these.[17]

This book explores a subset of biotechnologies, which I refer to as genetic technologies. These include various tools and techniques that were (and are) used in the knowledge that they cause physical changes in genes and chromosomes. Such technologies include means of generating so-called artificial mutations, changing the number of chromosomes, causing physical damage to or reconfiguration of chromosomes, and transferring genetic material between organisms or otherwise altering DNA using molecular genetics. One could argue, and people often do, that any method of creating change in the inherited traits of plants and animals—even unintentional selection such as the domestication of some crops by early humans—is a kind of genetic technology.[18] After all, breeding reshaped the genetic material of many species, sometimes very dramatically, long before genes and chromosomes were things that people thought about. Here, however, I distinguish between longstanding breeding methods such as selection and hybridization and more recent innovations like mutagenesis and transgenesis. I do so because most breeders, geneticists, and other observers of the twentieth century did so—and it is their world I seek to understand and explain. The production of "artificial" variation via physical tinkering with the material stuff of genes and chromosomes, whether achieved through radiation, chemicals, or other treatment, was seen as freeing breeders from the limitations of working with "natural" variation in heritable traits. These were means of speeding up evolutionary processes by bringing human ingenuity to bear on the very creation of such variation, a capacity that selection and hybridization did not seem to offer. A similar tale may be told of transgenic technologies. Although today many biotechnologists, facing widespread distrust and sometimes rejection of their efforts, are eager to show transgenic methods as little different from earlier innovations like mutagenesis or even traditional breeding, this was not the case when transgenic manipulation was first introduced. In the early 1970s, it was celebrated as a means of escaping evolutionary history altogether, enabling humans to cross otherwise impenetrable species boundaries and, through this creative recombination of genetic material, to innovate organisms that nature itself could not.

To better understand the history of this subset of biotechnologies—those explicitly conceived as tools for genetic manipulation—I reconstruct various efforts to produce such technologies in the mid-twentieth-century United States. Writing the history of genetic technologies as a succession of innovations aimed at the same goal, rather than tracing back to the origins of the transgenic approach we evaluate today as having been successful, offers a new perspective on this history.

Though they are no longer treated as the lone pioneers of effective biotechnologies, molecular biologists and their creation of tools for manipulating DNA continue to dominate the stories told about genetic technologies.[19] My narrative, which follows the lead of other historians, instead shows these biologists to be late arrivals to a lively world of research into options for manipulating genes or chromosomes and the collective imagination of what lifeforms might result.[20] In this book, molecular biologists appear only in the epilogue, as the champions of yet another potential technology for engineering life: recombinant DNA and the methods that it made possible. What becomes obvious in my account is the way in which these biologists' claims for what might be achieved with the tools and techniques of molecular biology echoed those of their predecessors working with mutagenic agents. Even more tellingly, their faith in the power of a brand-new technology mirrored the faith of earlier biologists in the cutting-edge technologies of their own decades, whether x-ray tubes, chemicals, or nuclear reactors.[21]

This account further reveals that the histories of biological technologies are far more interconnected with the histories of other technologies than our typical depictions of these have allowed, and not just in a shared enthusiasm for novel innovations. Although the history of biotechnology is a recognized part of the history of technological development (included, for example, in textbooks on that subject), this history tends to be presented as an independent thread in which biological researchers worked away at biological problems.[22] By comparison, I show how the development of new genetic technologies was interwoven with the work of other industries: electromechanical, chemical, and nuclear. I further argue that the pursuit of both innovations in breeding methods and innovations in biological forms traced the same patterns as the pursuit of innovations in these other realms, developing via an obsession with efficiency, in concert with many different users, and eventually in alignment with the self-perpetuating logic of large technological systems.

The result, an integrated history of biological research and technological development, provides an important bridge between our established histories of biology as technology in the twentieth century. It links the history of Mendelian genetics as a social and agricultural technology (which stretches from turn-of-the-century eugenic aspirations through the development of hybrid corn and then the Green Revolution) with that of the pursuit of molecular genetic technologies (extending, by some accounts, from a 1930s initiative of the Rockefeller Foundation on the "Science of Man" to the creation of transgenic organisms via molecular methods and beyond). It does so not by tracing the intellectual developments that led biologists to ever-greater insights into the molecular makeup of genes but via their dogged pursuit of tools—any tools from any area of technological innovation—that promised to alter the physical substance of genes and chromosomes, regardless of whether its mechanism or effects could be adequately explained by theory. Genetics research was always about innovation, of new biological tools and new biological types. And it was, without pause, productive of such tools and types.

To date, most histories have rightly stressed that the individuals and corporations that most vigorously pursued genetic technologies in the 1970s and 1980s hoped to profit from a growing global marketplace in biotechnology and were encouraged in the United States by friendly regulatory and intellectual property regimes.[23] Similarly, the story of agricultural biotechnology with which we are most familiar is one in which large corporations aggressively developed and promoted genetically modified (GM) crops; aided by intellectual property laws, they then used these GM varieties to gain control of an extraordinary share of the global seed market, even in the face of strong consumer resistance.[24] These accounts highlight the roles of scientists, corporate interests, and investors in spurring the continued development of technologies for manipulating genes.

The story told here is different. It is not solely about the scientists and researchers engaged in the development of genetic technologies, but includes many Americans who encountered these tools or information about them as farmers, gardeners, students, or simply as newspaper and magazine readers.[25] The aspiration for technologies that would grant control over the heredity and evolution of living organisms was not limited to the research programs of either quirky or mainstream biologists, to the ideologies of foundations and their grantees, or to the hopes of industrial producers and multinational corporations.[26] It was an aspiration and active pursuit shared among agriculturists, horticulturists, and many other Americans who believed that living things could be reshaped to human imagination provided only that the appropriate technologies were developed and perfected. Recent work by sociologists has shown that the "hope and hype" that surround certain innovations, and the articulation of future benefits to be derived from these, are crucial drivers of research and development. This is true whether these visions of the future are generated by individual researchers, the corporations that employ them, the governments that invest in or regulate them, an expectant public, or some combination of these.[27] As I will show, shared visions of a future of agricultural abundance made possible by innovations in genetic technologies spurred efforts to test, trial, and perfect these innovations—regardless of how wild the expectations might have seemed and how close they followed on the heels of disappointment with earlier efforts.

In the chapters that follow, I tell the history of Americans' encounters with early genetic technologies by considering three distinct moments of innovation in plant breeding: work with x-ray radiation in the 1920s and 1930s, with the chemical colchicine in the late 1930s and 1940s, and with radioisotopes and other nuclear technologies from 1945 until about 1960. Although I call attention in these chapters to a past in which there existed shared enthusiasm for the innovation of genetic technologies that would transform agriculture, I do not intend for this to indicate that the use of transgenic plants and animals in food production today is either safe or acceptable, or even that it has precedent.[28] In providing evidence that scientists and the public conceived of early genetic technologies as enabling, or soon enabling, them to engineer organisms to better meet agricultural needs, this book offers an argument about the aims and content of certain areas of scientific research and technological innovation in the middle decades of the twentieth century. And yet this perspective on the history of genetic engineering, and especially of public engagement with this area of research, does provide a fresh outlook on the subsequent trajectory of genetic technologies. Readers who are interested in what I have to say about contemporary genetic engineering and agricultural production should be sure to read the epilogue, where I treat this question at greater length.

Now, however, it is time to turn to the promised story.

PART ONE

Evolution by X-ray:
The Industrialization of Biological Innovation

In June 1928, some 150 "farm boys and girls" from across the United States gathered in Washington, DC, for the second-ever National 4-H Club Camp. During their five-day stay in the capital, these "junior farmers of America" were treated to talks and tours by various dignitaries and USDA officials in which they learned about topics ranging from the procedures of US policy-making to the work of the Department of Animal Husbandry. On the second-to-last evening of their stay, the campers sat to hear an address by Edwin Slosson, head of the news agency Science Service and de facto national authority on the latest in science and technology. Slosson offered up a rosy picture for these rural youngsters. According to a later news report, he declared that "agriculture is beginning upon a boom that will give it rank with the other great industries of the country." The boom would be made possible by science, which was helping make agriculture more efficient and more profitable: "Already it has turned many waste products of the farm into profits, and with the recent discovery that the X-ray can produce new and more profitable varieties of barley, wheat, and tobacco, the future outlook for agriculture is most promising."[1]

The 1920s were a difficult decade for most American farmers. While many people in the United States enjoyed a period of economic prosperity, farmers found themselves in straitened circumstances as a result of falling prices and shrinking demand for major crops. As industry and manufacturing boomed, and looked confidently toward the future, agriculture floundered, mired in the same vicious cycle of overproduction it had experienced many times before. Slosson's confident words aimed to assure the young farmers of the nation that this would not always be the case. Agriculture could be as successful as industry. The key was efficiency.

One route to increased efficiency appeared to be x-ray technology. The "discovery" to which Slosson referred was the 1927 demonstration that x-ray radiation could be used to create genetic mutations. Many people at the time assumed that such mutations, spontaneously occurring, were the basic stuff of evolutionary change, the source of the heritable variations found among all living organisms. When biologists announced that mutations could be produced on demand, simply by exposing a plant or animal to x-rays, the assumption was that scientists had found a way to intervene directly in evolutionary processes. Some even described the effect as "speeding up" evolution. One imagined use for this technique was the creation of better agricultural organisms, especially crop plants. If x-ray exposure generated heritable variation equivalent to that found in nature, then surely it could be used to create plants possessing a wide array of novel traits, some of which might prove useful in producing improved crop varieties. In other words, speeding up evolution might also mean speeding up agricultural breeding.

This goal—of making the development of new and useful products more fast-paced and predictable—was by no means novel. In fact, the same aspiration had underpinned one of the major developments in American industry in the first decades of the twentieth century: the rise of the industrial research laboratory. The General Electric (GE) Company established the first such laboratory in the United States in 1900; by 1927, there were nearly one thousand industrial research laboratories across the country.[2] In most cases, the central role of these laboratories was to provide the steady stream of new and improved products and processes that would keep the company ahead in the increasingly competitive industrial marketplace. Historians have described the development of industrial research laboratories as "the industrialization of invention."[3] Invention, a process once dominated by independent craftsmen and tinkerers working in their own workshops at their own pace, was now integrated into the industrial enterprise. A whole cadre of scientist and engineer employees, working under the direction of a laboratory manager, would presumably churn out much-needed inventions and further innovations more quickly and cheaply. That was the hope, at least.

A similar hope inspired sentiments such as Slosson's, that x-rays would make plant breeding faster and more efficient, a vision that could also be described as "the industrialization of invention"—except in this case, the sought-after inventions were not mechanical, chemical, or electrical, but biological. The process of creating or "inventing" a new plant variety depended on the independent, and unpredictable, variable of natural evolution. The appearance of novel traits, essential for producing crop plants with desirable features such as disease resistance or cold tolerance, occurred spontaneously (if at all), and the existence of these traits had to be discovered through surveys of known varieties. To some people, this seemed wildly inefficient. If, however, new traits could be produced regularly and predictably, and without the time and expense of scouring collections for them, perhaps the process of breeding a new plant variety would begin to resemble efficient industrial manufacturing rather than painstaking craftsmanship.

Perhaps then agriculture would gain rank with the other great industries of the country.

Mutation Theories

In all likelihood, the first person to voice a desire to control "mutation" in order to improve plant breeding, and to suggest radiation as a means of achieving this, was the very man who introduced the concept of mutation into biological theory: the Dutch botanist Hugo de Vries. In the summer of 1904, de Vries presented his ideas to an audience that had gathered in Cold Spring Harbor, New York, to celebrate the opening of the Carnegie Institution of Washington's Station for Experimental Evolution. By that time, de Vries was internationally known for his mutation theory in which he had proposed a new explanation of evolutionary change. He was enthusiastic about the research station at Cold Spring Harbor and the studies that were to be carried out there, believing it was crucial that researchers consider evolution using new approaches and new ideas. "We want to share in the work of evolution, since we partake of the fruit," de Vries explained, speaking of the mission of experimental evolutionists. "We want even to shape the work, in order to get still better fruits."[1]

To de Vries, the trajectory that experimental research on evolution should follow was clear. First, scientists would elucidate the nature of mutation, the basis of evolution, by working with species found to be changing in nature. Then they would seek the causes of these mutations. The culmination of the field would be the application of this knowledge to produce mutations—and therefore evolution—on demand: "New and unexpected species will then arise, and methods will be discovered which might be applied to garden plants and vegetables, and perhaps even to agricultural crops, in order to induce them to yield still more useful novelties." De Vries thought that the radiations generated by the curious phenomenon of x-rays and the recently discovered element radium might prove key to the final part of this research program in which scientists would generate mutations at will.[2]

When de Vries referred to "mutations," he did not have in mind the biological events this term would later be used to describe. In his mutation theory, first published in 1900, he used the word to indicate the striking changes in inherited traits he observed among his experimental flowers, changes that he believed set one species apart from another in a single generation. He considered these leaps in form to be a more important component of evolutionary change than the Darwinian process of gradual differentiation via natural selection. De Vries's theory rose quickly to prominence, inspiring all kinds of research into mutation. Although it fell out of favor almost as fast as it had risen, the notion of mutation found a lasting place in biological theory. What's more, many biologists in subsequent decades would share de Vries's vision of evolution as a process that could potentially be directed by human beings through their control of mutation.

Before turning to that history, it is essential to place de Vries's ideas about the study of evolution and heredity in a broader social and intellectual context. His emphasis on the literal fruits of research in his 1904 address, for example, is a reminder of the immense importance attached to the potential practical payoffs of the study of heredity and evolution at the turn of the century. This point will be obvious to readers who know the story of the so-called rediscovery of Gregor Mendel and the subsequent reception of his studies of inheritance in sweet peas: Beginning in 1856, Mendel, an Augustinian friar in Brno, Moravia (then part of the Austrian Empire), hybridized plants and tracked the inheritance of specific traits from generation to generation. In 1865, he reported his results to his local natural history society. He described how the visible characteristics of his pea plants—such as their height, the color and texture of their seeds—were determined by discrete hereditary elements within their cells. Mendel's observations and statistical analyses of the patterns of inheritance enabled him to propose a set of rules that governed the behavior of the hereditary elements, later called Mendelian factors or unit characters, and therefore the appearance of the traits they determined. These rules in turn could be used to predict the distribution of traits among the offspring of a given hybrid combination of pea plants. This work drew scant attention until a few scientists, de Vries included, came across it in the course of their own research around 1900. New generalizations about inheritance based on Mendel's ideas, generalizations that we know today as the laws of segregation and independent assortment and the concept of dominant traits, soon circulated widely, welcomed especially by many biologists, breeders, and eugenic advocates.[3] It is not hard to see why. The laws seemed to offer both an explanation for patterns long observed by breeders and hybridizers—and therefore a validation of their methods—and a route to predicting and potentially controlling the inheritance of traits in plants, animals, and humans alike. Mendel's ideas found many champions, especially in the United States and Britain. Their "rediscovery" is generally credited with sparking the development of the field we know as genetics.[4]

In the United States, the promise that heredity might be better understood and controlled meshed well with ambitions for a more scientific approach to agriculture. The scientific investigation of inheritance promised to be economically relevant, as research in the emergent discipline of genetics could easily be directed toward agricultural improvement. Studying patterns of inheritance in chickens, for example, might produce better egg layers as much as it might illuminate the underlying mechanisms through which traits are passed on from one generation to the next. And just as researchers pursuing studies of Mendelian inheritance sought to show how their knowledge might be applied in this way, breeders sought to demonstrate how the new genetic ideas aligned with their existing expertise and established methods. The upshot was an alliance between advocates of Mendelian genetics and agricultural breeders, one that helped ensure the success of Mendelism in the United States, in laboratory and farm fields alike.[5]

From about 1900, then, the introduction of Mendelian genetics provided a conceptual and practical tool for American breeders to claim control over the direction of evolution in domesticated plants and animals. Yet Mendel's laws neither explained the origins of variation in heritable characters among individuals or types nor indicated how such variations might be produced—and variation was the crucial currency of both natural evolution and selective breeding. One could hardly imagine a new species or an improved breed coming into existence without some characteristics that set it apart from its progenitors. As a result, ideas beyond Mendelian genetics proved influential in spurring early research into the nature of heritable variation. Foremost among these ideas were de Vries's mutation theory of evolution, with its emphasis on understanding and perhaps controlling the appearance of novel types, and the work of his contemporary Luther Burbank, an American horticulturist, in creating new varieties of flowers, fruits, and vegetables. Each in its own way encouraged people to believe that the variations in heritable traits found among plants and animals might be generated on demand.

Hugo de Vries's botanical investigations began early. He had been an avid botanizer even from childhood, gathering specimens from the countryside near his family home in Haarlem, and his interest in the study of plants persisted into adulthood. After graduating from Leiden University, the young de Vries spent a few summers in the early 1870s studying plant physiology under the eminent German botanist Julius von Sachs before going on to establish an independent research career. While at university, de Vries had read Darwin's *On the Origin of Species* and had become something of a Darwin devotee, an interest that inspired his turn to the study of heredity and evolution in the late 1870s. Hoping to get a better grasp of these processes, de Vries carried out extensive plant hybridization experiments. It was this work that led him to the 1865 paper by Mendel on inheritance in sweet peas, and subsequently to its recirculation and then celebration.[6]

Having been instrumental in calling attention to the theories of Mendel, de Vries followed up with a novel biological theory of his own. In *Die Mutations-theorie*, released in two volumes in 1901 and 1903, de Vries sought to revise one of the central tenets of Darwin's evolutionary theory, the notion that evolution and speciation occur through gradual change.[7] Darwin's hypothesis was that natural selection acted on the small differences among individuals—so-called continuous variations in traits—to produce evolutionary change over many generations. De Vries disagreed. According to his mutation theory, new species were assumed to arise as a result of sudden and distinct changes in form, also called discontinuous variations or "saltations." Natural selection might then act on these saltations to weed out the less fit of any new "elementary species."[8]

De Vries had arrived at this conclusion through his observations of the evening primrose, *Oenothera lamarckiana*. In 1886, he had encountered a population of these flowering plants growing in a field near Amsterdam and had been struck by the species' unusual displays of variability. Keen to observe the plants more closely, he transplanted several to his experimental garden and cultivated them over a few generations. The plants rewarded his curiosity. Instead of offspring looking for the most part like their parents, generation after generation, as one might expect, *Oenothera* plants were prone to producing offspring that differed in basic characteristics (fig. 2). De Vries introduced the term "mutation" to describe these deviations in form. He inferred from his observations that he was watching speciation in action: the differences from one generation to the next were, in his estimation, significant enough to mark the origin of an entirely new species of *Oenothera*. The further implication was that species might well arise from abrupt alterations in form, as opposed to small gradual changes accumulating over time as Darwin had proposed.[9]

Although de Vries drew his strongest evidence from *Oenothera*, he also looked to the accumulated wisdom of horticulturalists, especially those working with flowers and fruits, to support his mutation theory of evolution. Horticulturists were familiar with the sudden occurrence of distinct and stable new forms, such as a white rose with an unusual flush of pink or a peach with

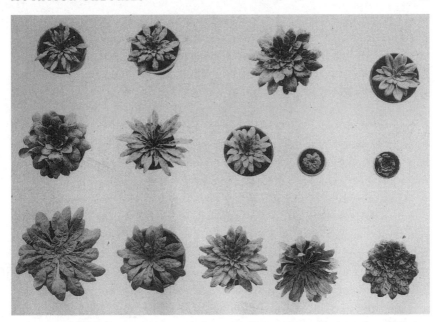

FIGURE 2. Hugo de Vries's encounters with the many variations of *Oenothera lamarckiana* led him to his mutation theory. Here the parental form of the plant in the lower-right corner is compared to eleven "mutants"; at upper left, an "aberrant" of *Oenothera biennis* is compared to the "typical rosette" of the same. From D. T. MacDougal, A. M. Vail, and G. H. Shull, *Mutations, Variations, and Relationships of the Oenotheras* (Washington, DC: Carnegie Institution of Washington, 1907), plate 1.

smooth skin rather than fuzzy. These were referred to as "sports," and those that were particularly appealing could be propagated through grafting, cutting, or other means, and sold as novel varieties. To de Vries, "the so-called sports," these sudden leaps in form used to establish distinct lines, were the best-known examples of what he now categorized as mutations.[10]

De Vries's mutation theory proved influential in the first decade of the twentieth century. Far more important than its challenge to Darwinian evolutionary theory, however, was its effect on methods of biological investigation, as de Vries's speculations spurred other scientists to study evolution via experiments.[11] For example, Thomas Morgan, a professor of zoology at Columbia University in New York, began searching in 1907 for a way to study evolution experimentally just as de Vries had envisioned. Working with the common fruit fly, *Drosophila melanogaster*, Morgan initially sought to produce or discover a de Vriesian mutation. His efforts, slow to start, eventually led him to identify many distinctive and apparently new traits among his laboratory populations. This in turn led him into what would become an enormously productive and influential experimental program—though not in experimental evolution but in genetics.[12] De Vries's theory was also of great interest to those engaged in plant breeding. Just as the theory sought to explain the origin of new species, so too did it shed light on the production of new cultivated varieties. As de Vries suggested, an appreciation of the "high practical value of the elementary species [those newly arisen via mutation] . . . will, no doubt, soon change the whole aspect of agricultural plant breeding."[13] Journalists proved particularly responsive to this facet of de Vries's work. Sensational news reports, inspired by lectures given by de Vries or his admirers and running under headlines like "How to Increase World's Foods" or "Grow Larger Grain," declared that mutation theory would soon lead to higher-yielding rice and wheat.[14]

The application of mutation theory to breeding practices was hardly as direct as these reports, or de Vries himself, suggested. At the time, most agricultural breeders working with field crops grown from seed—oats, barley, corn, cotton, and others—worked to improve varieties through mass selection. Breeders would save the seeds from many of the "best" plants grown in a season, judging "best" by whatever criteria they thought most appropriate, and sow this mixed lot of seed the following year. The process was continuous. A breeder had to attend season after season to the most desirable traits or else the population would gradually return to its original form. How would mutation theory improve on this practice? For one, de Vries reasoned that breeders could be saved the trouble of selection, if only they could be taught to seek out plants bearing true mutations. These would not revert to ancestral characteristics if left to breed freely in the way that varieties maintained through selection of normal variations would. In other words, looking for desirable mutations and selecting these would make breeding far more efficient. But de Vries was not content to simply observe and utilize mutations where these had arisen through natural processes. He also hoped that the appearance of mutations might be controlled. "A knowledge of the laws of mutation must sooner or later lead to the possibility of inducing mutations at will and so of originating perfectly new characters in animals and plants," he speculated. This in turn would "place in our hands the power of originating permanently improved species."[15] Once this power was achieved, breeders would be spared waiting and searching for spontaneous changes when seeking desired traits, just as they would be spared the task of continuous selection. In this vision, mutation theory, and mutation research, would revolutionize breeding practices.

To de Vries, the ability to direct mutation—and with it, evolution—was a thing of the future. But others at the time believed that skilled breeders already possessed this ability. Not long after the publication of his mutation theory, de Vries became embroiled in a high-profile dispute on exactly this point. At the center of the commotion was a famous California horticulturist named

Luther Burbank. In the decade before the rediscovery of Mendel and de Vries's proposal of his mutation theory, Burbank had captured national and even international attention through his creation of impressive new varieties of flowers, crops, and especially fruits—and through his canny self-promotion. By the turn of the century, he was a much-loved public figure. Just as there was the "wizard of Menlo Park," the great inventor of mechanical devices Thomas Edison, there was a "wizard of Santa Rosa," the inventor of flowers and fruit trees Luther Burbank.[16]

Early in his career, Burbank had garnered a reputation for creating striking improvements in fruit tree varieties. In one famous example of his efforts, Burbank in the late 1870s and 1880s developed several varieties of orchard plums. In order to introduce novel characteristics into the varieties already established in the United States, Burbank imported Japanese plum varieties, which he then hybridized with American and European ones. By 1887, he had forty-three hybrid varieties ready to share, many displaying new qualities derived from the Japanese plums, features like red flesh and earlier ripening. These proved to be an immediate success and counted among the most popular plant innovations he ever produced.[17] The process by which Burbank had produced the plums was typical of his methods. He sought out desirable traits in specific plants or varieties and bred these with individuals from the stock he wanted to improve, selecting for the coveted traits in subsequent generations.

Though his methods were in practice straightforward, in his writings and lectures Burbank often described his work in almost mystical terms, and rarely with reference to his specific practices or current scientific ideas. This in turn provoked confusion over how exactly he produced his results and how much influence he actually had over plant heredity. Was he simply making use of heritable variations found in nature, or was he creating these through his work? Was he a plant breeder or a magician?[18] Journalists, promoters, and Burbank's many early biographers tended to think the latter. In the press, Burbank was frequently depicted as a wizard conjuring up new forms of life.[19] After all, he did not seem to be limited in his work by what could be found in nature. As one journalist reported in the 1890s, "Mr. Burbank has proved that . . . every form can be made to 'break,' no odds how seemingly obdurate it may seem, and when once 'broken,' it may be carried in any direction at will by time, skill, and patience."[20] In the late nineteenth century, the term "break" referred to the appearance of a new form or trait among plants of an established variety, such as a plant with red flowers where ordinarily white was expected or an unusually short plant in a variety known for height. To suggest that Burbank could make any plant "break" was to suggest that every plant could be forced to show variation in the hands of a capable breeder.

The publication of de Vries's mutation theory offered an opportunity for Americans to reconsider Burbank's work in light of new scientific research—or to use the example of Burbank to discredit de Vries's ideas. In a particularly generous biography of Burbank published in 1905, the journalist William Harwood declared mutation theory "to have been overthrown by Mr. Burbank." De Vries had assumed that nature governed the appearance of mutations, but according to Harwood, "Mr. Burbank, times without number, has produced these strange mutations at will. They can be produced, he says, by anybody who systematically sets to work to disturb the life habits of the plants." Harwood further suggested that Burbank's successes could be imitated by nearly anyone working in the field or garden.[21] By this logic, de Vries's claim that someday scientists might discover how to control the process of mutation was ridiculous. This knowledge was already in hand.

These sensational reports about Burbank caused de Vries to wonder whether the Californian was in fact producing mutations. Intending to answer the question for himself, de Vries visited Burbank's farm. The observations he made on a first visit in 1904 confirmed for him that Burbank's work was in fact quite ordinary. De Vries had wanted, for example, to determine how Burbank had created a pitless prune, a creation that surely represented the radical alteration of a standard variety. What he discovered in his interview was that Burbank had simply imported a pitless variety grown in France and crossed it with some of his own varieties. He had not originated the character, a feat that would suggest he had methods of controlling mutation, so much as skillfully incorporated it.[22]

When an initial effort to explicate Burbank's methods failed to quell debates, de Vries tried again, and more vehemently, to convince observers that Burbank had no extraordinary power over mutation. He even made another visit to Burbank's farm to gather evidence. If Burbank could produce mutations, de Vries argued, then he "would probably be able to teach me to do so, too. The work on the causes which provoke mutations, which is still before me, could be materially shortened by such lessons." In 1906, Burbank remained confident of his abilities, reportedly telling de Vries, "I can produce them [sports] by the hundreds and at will."[23] But de Vries went on to argue that Burbank's sports were not true mutations as defined in the mutation theory. They were simply novel combinations generated through hybridization, or variations in traits stabilized through selection over a number of generations. Only in a few circumscribed cases, such as that of "bud sports" on fruit trees—branches bearing distinct foliage, fruit, or other features—did Burbank's sports constitute de Vriesian mutations, and as de Vries concluded, these were very rare, even for the horticultural wizard Burbank.[24]

De Vries had his own ideas about the means by which mutations could potentially be generated, and they bore little resemblance to any of Burbank's activities. One idea, presented at Cold Spring Harbor in 1904, was to use x-rays and radium. De Vries would hardly have needed to explain this choice to his audience. X-rays, discovered by Wilhelm Röntgen in 1895, and emanations of radium, first identified by Marie Curie in 1898, both excited immense scientific and popular interest at the turn of the century and through the 1910s and 1920s.[25] These were mysterious forces, and either might be capable of achieving an influence over living organisms previously thought impossible. Just ten years after their discovery, x-rays were used in biological, physical, and chemical experimentation; they were deployed in medicine as a diagnostic tool and as a therapeutic; they were incorporated into sideshows and sales pitches.[26] The discovery of radium emanations similarly informed new theories and experiments, with many scientists viewing the "living atom" as an especially suitable tool for interrogating the nature of life.[27] Imbued with many powers, radium insinuated itself into treatments and tonics, as well as consumer goods like illuminated clocks and watch dials.[28] Radiation, then, was already well established as a potentially transformative tool.

It was not long before experimenters followed up on de Vries's proposals. His mutation theory inspired many investigations, including studies that used the novel radiations alongside other potent substances in order to induce the appearance of mutations. In a series of experiments begun in 1905, the botanist Daniel MacDougal of the New York Botanical Garden injected various chemical compounds into the ovaries of *Oenothera* plants before fertilization in hopes of producing changes in the characteristics they passed on to their offspring. Though the experimental procedure had been derived from similar experiments conducted by Charles Darwin, MacDougal had been drawn to the study of mutation by his reading of de Vries's work. In December of that year, he presented the results of these experiments, claiming to have been successful in the endeavor and therefore the first experimenter to cause a mutation to occur through direct stimulus.[29] MacDougal subsequently turned to the use of radium to determine whether this would have similar effects, and he soon claimed another success: Radium, too, could bring about mutations.[30] Although many biologists accepted the veracity of these achievements, not everyone was in agreement. Among the most notable skeptics was MacDougal's collaborator at the New York Botanical Garden, C. Stuart Gager, who carried out more extensive general investigations into the effects of radiation on plants during the same period. Gager wondered whether the novel types that appeared in radium-treated cultures could accurately be labeled new species—a condition necessary for them to be true de Vriesian mutants.[31]

In conducting his experiments, MacDougal was especially interested in the exploration of de Vries's mutation theory and the notion of experimental evolution, and his methods reflected a broader trend in which biologists were increasingly drawn to experimentation as a mode of research. No doubt many of his peers saw this research as contributing primarily to debates on mutation and the origin of species.[32] But just as de Vries's mutation theory could be imagined (by de Vries as well as by his audiences) as the likely source of an agricultural revolution, so too could MacDougal's work be interpreted in this way. A 1907 essay in the American monthly magazine *World's Work* described breeders' interest in methods that would "shake up the protoplasm of any plant or animal," and offered evidence that such methods in fact already existed by describing MacDougal's experiments on *Oenothera* along with those of the biologist William Tower on mutations in the potato beetle. The further development of these methods, the author maintained, would ultimately lead to significant advances in plant breeding. The time to the production of a variety would drop from many generations to just three weeks. Not even the wizard Burbank could compete with such a method: as the report concluded, whereas a potato that had been developed by Luther Burbank (the Russet-Burbank, a chance find that became his earliest moneymaker) "has enriched the country by more than $17,000,000," the "Tower-MacDougal method" might make "even Burbank's earliest triumph seem very literally 'small potatoes'!"[33]

Such ideas and enthusiasm existed outside the hyperbolic science reporting of the American press as well. MacDougal's work inspired a few biologists and breeders to think it might be possible to control the process of mutation and, in doing so, to improve agricultural production. John Coulter, head of the Botany Department of the University of Chicago, offered one such assessment. He took MacDougal's findings to indicate that artificial production of mutations through the "'disturbance' of the germ plasm" might be achieved, and he further anticipated that when "the conditions that favor mutation are known and under control, mutations may be multiplied artificially and desirable new forms secured in this way."[34]

It is clear that aspirations for tools and methods that would generate mutations on demand accompanied the very invention of mutation as a concept describing biological change, as did an understanding that the ability to induce mutation would be a likely windfall for agricultural production. Although de Vries's mutation theory of evolution soon fell out of favor, and MacDougal's pioneering experiments were eventually forgotten, interest in mutation persisted—and the notion that radiation might provide a means to disrupt heredity continued to inform scientists' varied efforts to reach this goal.

An Unsolved Problem

Interest in the potential agricultural and horticultural payoffs of human-induced mutations abated somewhat during the later 1910s and into the 1920s, at least in comparison with the heady statements that followed on from de Vries's speculations in 1904 and MacDougal's claims a few years later. There are two likely explanations for this shift in the prevailing winds. First, de Vries's mutation theory came to be increasingly discredited, especially after further studies of *Oenothera* revealed that the aberrant forms possessed unusual chromosome arrangements that in turn accounted for their unusual patterns of heredity. Second, plant improvement methods—and other areas of genetics research—came to focus more intensely on the development, through inbreeding, of pure lines.

The latter shift entailed a striking change in breeding practices. For example, pure-line breeding of corn and the subsequent crossing of inbred lines to create vigorous hybrids, techniques first advanced only after the turn of the twentieth-century, increasingly dominated breeding efforts at US agricultural experiment stations by the 1920s.[1] In establishing purebred lines, breeders aimed to generate a variety with known characters that would appear reliably from generation to generation. The method of producing a pure line was similar to breeding via the search for mutations in that both eschewed the practice of mass selection in favor of the selection of superior individuals. Yet in its attention to reducing variation in inheritable characteristics, it was almost the exact opposite of the as-yet-only-imagined process of improving plants via induced mutation. Attention to the creation of ever-more-uniform and ever-more-predictable crop varieties would dominate breeders' activities for much of the twentieth century.

Still, aspirations for a method that would create inheritable variation on demand did not disappear. Some breeders saw the task of producing variation in important traits as a central concern. And the possible artificial production of mutations also continued to be a goal of experimental researchers, geneticists in particular.

The first two decades of the twentieth century saw the discipline of genetics take shape as a distinct field of study. Geneticists aimed to elucidate the features of inheritance common to all living things even as they narrowed their focus to the genes and chromosomes of a few chosen model organisms such as the *Drosophila* fruit fly or *Zea mays*, better known as corn or (as geneticists preferred) maize. In these organisms, mutations—a term that by the 1910s increasingly referred to heritable changes in Mendelian factors or unit characters, traits that were inherited in typical Mendelian patterns—were crucial tools for visualizing the otherwise invisible genes and tracing their inheritance over multiple generations.[2] Geneticists eagerly searched for these among their laboratory and field populations. They also sought means of producing such mutations on demand. In their 1925 *Principles of Genetics*, the geneticists Edmund Sinnott and L. C. Dunn discussed recent experiments testing x-rays, radium, and chemical injections as means to produce heritable changes. None had proved satisfactorily (or at least, to universal agreement) whether and how mutations could be consistently induced. In Sinnott and Dunn's estimation, the origin of mutation was "one of the most interesting unsolved problems of genetics." And this was not just a matter of understanding how exactly heredity worked. They later underscored again the central importance of understanding mutation—this time in relation to the task of plant or animal improvement. As Sinnott and Dunn described, the fundamental challenge facing a breeder was how to control the inheritance of traits. "This control," they argued, "presupposes the power not only to manipulate inheritance by breeding operations, but also the power consciously to induce new heritable variations."[3] In other words, even as genetics expanded to encompass many distinct phenomena and an array of methods and organisms, its practitioners continued to view the ability to induce mutation as a potentially transformative capacity.

Although many biologists considered the origin and nature of mutation a central problem for understanding inheritance, as well as an important element of evolutionary change, it was but one topic of research among many that interested geneticists during the early 1920s—and only one imagined route to altering patterns of inheritance. Studies of phenomena known as linkage and crossing over, for example, were a far more common feature of early genetics research, and similarly understood as pathways to achieving greater human control of heredity and evolution.

By 1920, geneticists had determined that for any given species there were a defined number of linkage groups within which Mendelian unit characters tended to be inherited together, and these quickly became a central focus of research. Within these linkage groups, which came to be associated with specific chromosomes, characters displayed a range of degrees of linkage: some appeared together in generation after generation without change, others displayed only a very loose association, and the rest fell all along the spectrum between these two extremes. The explanation for this range in frequency of linkage was the occurrence of crossing over between chromosomes, an event believed to occur during the development of sperm and egg cells. In crossing over, a segment of one chromosome is swapped with the corresponding segment of the homologous chromosome; this conceivably would lead to the disassociation of two characters, depending on how far apart they lie on the chromosome.[4] *Drosophila* researchers working under Thomas Morgan at Columbia pioneered the use of data on the frequency of linkage (and therefore crossing over) to generate what became known as "maps" of the order and relative distance of factors—or genes, as they were increasingly known—on a chromosome.[5] These maps were a central tool of genetic analysis among the *Drosophila* researchers throughout the 1910s and 1920s, and inspired mapping projects in other organisms, most notably maize. Attention to mapping as a means of visualizing the otherwise invisible genes kept research on linkage and crossing over at the forefront of genetics research.[6]

One of the *Drosophila* geneticists' earliest discoveries about linkage was that the rates at which specific characters were inherited as a pair, though often remarkably constant, sometimes fluctuated. Calvin Bridges, a member of Morgan's fly group, first identified such fluctuations in the fruit fly in 1915 when he demonstrated that the frequency of crossing over for at least one region of a *Drosophila* chromosome was dependent on age.[7] This discovery encouraged geneticists to see whether patterns of linkage could be altered by other means. Like induced mutation, affecting linkage relationships by some external means suggested a way of disrupting expected patterns of inheritance. *Drosophila* researchers working on this problem soon demonstrated that the rate of crossing over could be influenced by the application of high temperatures and by selection over multiple generations—that is, that the process was governed by both environmental and genetic factors.[8]

One person who took a keen interest in this area of research was Willis Whitney, head of the General Electric Company's in-house research laboratory located at GE headquarters in Schenectady, New York. Whitney's employees—expected to be the source of important scientific and technological developments for GE—had been responsible for major innovations in x-ray technologies in the 1910s. These included, most notably, the physicist and engineer William Coolidge's invention of the high-vacuum x-ray tube in 1913. The "Coolidge tube" offered increased accuracy and intensity over existing x-ray technologies and eventually paved the way for more widespread use of x-rays by physicians and scientists as well as in industry. The patented device also offered GE a chance to leap ahead in the x-ray marketplace and spurred further x-ray research at the laboratory.[9] Although the main task for GE employees in this regard was the continued development of x-ray technologies, Whitney also hoped his staff could discover more about the biological effects of x-rays. Other accounts suggest that he was interested in not only the potential medical uses of the devices (which would undoubtedly bring greater profits to GE) but also their potential hazards.[10] With these interests in mind, Whitney engaged a professor at nearby Union College, James Mavor, to conduct studies of the effects of x-ray radiation on heredity, including effects on linkage and crossing over, using the *Drosophila* fly.

Mavor's experiments, conducted with the assistance of GE technicians, suggested that x-rays did in fact disrupt expected patterns of heredity. Radiation appeared to alter the chromosomes themselves, changing the observed rate of crossing over and causing nondisjunction (i.e., the failure of a chromosome pair to separate during cell division) in some cases, and also influenced fertility and sex ratios in the subsequent generation of flies.[11] Mavor was optimistic about the future of this research, predicting that scientists would search either for a chemical or for a particular physical condition, such as temperature or pressure changes or being hit by a beam of radiation, that could target a single gene or group of these, leaving the rest of the cell unharmed.[12] Whitney also found the results encouraging, describing to a reporter in 1924 how "the X-ray can modify the transmission of certain characteristics in the fruit fly," and promising that GE would further investigate the phenomenon.[13]

These results did not make the splash among genetics researchers for which Mavor—and likely Whitney—might have hoped; they were barely referenced before 1927.[14] One scientist who did consider the work to be at least of interest was Lewis Stadler, who was then exploring means to disrupt linkage in his own preferred experimental organism, maize. The pieces published by Mavor made it into his pile of reading material, and x-rays soon appeared on his list of experimental treatments.[15]

In 1923, Stadler was an employee of the Department of Field Crops at the University of Missouri. He had trained there as an agronomist and had been hired to the permanent staff in 1920. His work initially encompassed many practical demands of the experiment station, such as testing new varieties and corresponding with local farmers, but he developed an interest in genetics that soon absorbed most of his attention. Later biographical accounts attribute this shift to his having read Morgan's 1919 synthesis of genetics research, *The Physical Basis of Heredity*, but it is hardly surprising that a researcher at an experiment station in the 1920s, heavily invested in the improvement of crops, would become interested in the burgeoning field of genetics.[16] Across the country, agricultural station researchers in positions like Stadler's were taking up genetics, work that spoke to both academic-scientific and practical farming communities. Researchers at the Missouri station were redirecting their efforts along just such lines, having begun a project on maize genetics in 1919.[17]

Maize, much like *Drosophila*, was a common subject of early genetics research. Geneticists working with maize studied the inheritance of quantitative characters (such as the number of rows on an ear) and qualitative characters (such as color and texture), considered variations in linkage, and undertook cytological analyses—as the *Drosophila* geneticists did for the fruit fly.[18] Part of the appeal, and a key reason that maize research gained early popularity, was its association with practical needs and interests. Working with an experimental organism that was also the most economically important crop in the United States helped justify experimental genetics research at government-funded agricultural stations. One could hardly hope to improve corn scientifically without a clear understanding of how its inheritance worked, or so researchers claimed.[19]

In 1923, Stadler began a study of variations in rates of linkage and crossing over. In subsequent journal articles, he argued that a study of this topic constituted a novel contribution to genetics research. Despite the fact that *Drosophila* researchers had documented cases of variation in linkage intensity resulting from both heredity and environment, there were no equivalent published studies in plants. Moreover, as Stadler pointed out with reference to several specific experimental advantages, maize would allow him "to investigate certain phases of the problem for which the plant is well suited and the fly is not."[20] In correspondence, Stadler suggested still other reasons why the work was significant. In 1923, the production of inbred lines (i.e., lines that had been self-fertilized over a number of generations to ensure the inheritance of specific traits) was increasingly central to maize improvement. The maize geneticist and corn breeder Donald Jones had proposed the double-cross technique for hybrid corn, which involved the crossing of four unique inbred lines, in 1918, and the method was soon championed as a surefire route to higher-yielding varieties.[21] The production of hybrid corn, and consequently also the practice of inbreeding, became popular among experiment station workers in the following decades. It was obvious to Stadler that inbreeding could be used to eliminate undesirable traits. "But of course," he wrote to a colleague, "due to linkage we may eliminate a great deal of desirable germ plasm as well when desirable genes are linked with those which are eliminated." Wouldn't it be useful, then, if linkage rates could be altered? Stadler indicated that he was anticipating experimental results in this very area.[22] He no doubt saw that his study, if successful, might suggest means for breeders to disassociate traits normally inherited together. This in turn would give breeders a greater number of alternatives in terms of individual plants with different combinations of characters to use in their breeding programs.

Stadler conducted a preliminary round of experiments in the summer of 1923 in which he tracked differences in the linkage of three traits in different strains of maize under varying environmental conditions. Encouraged by the results of these initial trials, he planned a more extensive investigation for the following season. This would include different strains of corn, successive plantings, two or more ears of each of one hundred plants, and observation of crossing over in both male and female gametes.[23] He also decided to test a number of "environmental conditions" unlikely to be found in an ordinary cornfield. These included treatments with chloroform and chloral hydrate and exposure to x-rays and ultraviolet light, experimental variables likely inspired by those used in *Drosophila* experiments. He would apply these and other treatments to potted corn plants at three different stages of their growth, in "an effort to modify linkage relations."[24] The x-ray exposure in particular required a little extra legwork, necessitating that he transport several potted maize plants to a nearby hospital where he had arranged access to an x-ray machine.[25]

The outcome of the x-ray treatment was not entirely satisfactory. As a result of having been grown in small pots, the treated maize plants were "small and weak," producing no ears and very little pollen. Still, he managed to use the x-rayed plants in a series of pollinations, the progeny of which suggested that the radiation treatment had increased crossing over.[26] Though the experiment was not entirely satisfactory, the changes Stadler had seen in the treated maize plants led him to believe that he was closing in on an important discovery. As he explained to a representative of the Victor X-Ray Corporation of Chicago, from which he hoped to rent a portable x-ray machine so that he could irradiate plants as they grew in the field, "The object of these experiments is to devise methods for affecting heredity by external treatments. This is a problem of fundamental importance in biological science and has very important applications in animal breeding and plant breeding." He concluded, "We have good reason to believe that X-ray treatments may accomplish such changes."[27]

Stadler's unusual request to rent a portable x-ray machine for just a couple of weeks—for outdoor use, no less—was eventually approved. Many of the experimental treatments that he undertook with this machine in the summer of 1925 aimed to influence heredity by disrupting expected patterns of linkage.[28] To carry out these experiments, he wheeled the machine out into the open field on several occasions over the course of two weeks, irradiating for five to eight minutes at a time the pollen-producing tassels of forty individual corn plants as they matured.[29] Stadler then gathered pollen from the treated tassels and untreated controls and used these to pollinate the ears growing below. The kernels that developed from this mating would provide him with evidence as to the genetic effects of the x-ray treatment. The results (tabulated after the growing season had ended and the ears had been harvested) did indicate to Stadler that there might be an effect of higher radiation. Unfortunately, the natural variation in crossing over among plants rendered the study inconclusive. A second and slightly modified trial, carried out in a subsequent season, would prove similarly unhelpful.[30]

This outcome might have been disappointing, but in 1925 Stadler's attention was already turning to new questions. That summer saw the start of what he described as an "intensive study of variation in the rate of mutation" and the effects of different factors on that rate.[31] He also appears to have begun a study of the effect of x-rays on the appearance of mosaic endosperm, a condition in which some individual kernels of maize show both the dominant and recessive of a particular character.[32] And there was one result of the 1925 experiments that was clearly of surpassing interest to Stadler, a certain individual maize plant that he enjoyed showing off to friends and visitors at the station. It was a plant in which he believed he had caused a mutation to appear.[33] Although the experimental origins of that particular plant are unclear—one would like to assume that it had been the result of the x-ray studies, perhaps even an x-ray-induced occurrence of mosaic endosperm, which would explain Stadler's aggressive pursuit of that outcome—his appreciation of the significance of producing mutation is clear enough. He soon turned his attention to the question of x-ray-induced mutation, tackling the problem of how to influence heredity via this different route.

Stadler evidently had reason to believe from his studies that mutations could be produced through human intervention, but linking an observed phenotypic change to an alteration in a gene or a chromosome and then to the effect of x-rays was not easy. A geneticist hoping to demonstrate this link first had to start with known purebred lines, lest an observed mutation be simply an ancestral trait newly revealed. Then generational studies would be needed, to prove that any observed change was actually inherited, in Mendelian ratios, and therefore a germinal change. Of greater consequence, mutations were notoriously infrequent in their occurrence; as a result, unless the rate was drastically increased, or massive numbers of individuals evaluated, it was difficult to prove that a few observed changes were statistically significant. These were the challenges that other researchers working on induced mutation faced in making strong claims about the apparent genetic changes they discovered or induced in their experimental plants and animals. Stadler faced them, too. In January 1926, while continuing his work on maize, he began to plan an experiment that would take advantage of biological characteristics of the barley plant to show the effects of x-ray and radium radiation on the frequency of mutation.[34] In doing so, he aimed to meet the challenges of demonstrating that the effect caused by exposure to x-rays was in fact an alteration in a gene.[35]

For the barley experiment, Stadler borrowed time on an x-ray machine at the University of Missouri's Department of Physics and a supply of radium from a physician at Columbia University. He then exposed germinating seeds to radiation, for a few minutes in the case of those treated with the x-ray machine and several hours for the radium-treated seeds. One can only imagine that it was to Stadler's delight that after the first round of treatment, he observed irregular traits in both the radium-treated plants and the x-ray-treated plants. These were unusual chlorophyll types that would form the basis of his first public claims to having induced mutation. Among the x-ray-treated plants, Stadler discovered 14 "mutant seedling characters" out of about 1,200 flower heads; among the radium-treated plants, he found 3 in about 1,000 flower heads. By comparison, he could locate no mutants at all among the 1,300 controls. The induced characters that Stadler identified included three chlorophyll mutations, two of which had previously been seen in barley and were known Mendelian traits. Subsequent generations grown from self-fertilized seed confirmed that these were both hereditary and passed on in the expected Mendelian patterns.[36]

The work was, by all indications, progressing well. Stadler enthusiastically reported to the USDA in June 1927 that the corn investigations, which included both breeding work and the maize genetics study, were amply funded. In addition, his department had finished some construction over the winter months that would greatly assist him in his research. "We built a new greenhouse and a convenient little shack alongside the corn field for cytological and x-ray work (with water and electrical connections)," he noted. And circumstances seemed likely to improve still further. "We have a scheme up now to get an X-ray outfit permanently in place of the one we have been renting during the summers," he wrote.[37] He would be able to work up further confirmatory findings with barley, and perhaps expand the mutation studies to maize, more easily than ever. Stadler knew just how important this area of research might prove to be and how close he was to a major breakthrough. His colleagues and advisers in maize genetics agreed about the significance of the work, a fact that is not surprising given the longstanding interest among geneticists in finding means of disrupting heredity by any possible means.[38]

But others were closing in on a solution to this unsolved problem, too. At nearly the same time that Stadler was securing proof that mutation could be induced through radiation treatment, another research team in California was excited to discover similar effects. The botanist Thomas Harper Goodspeed and the chemist Axel Olson, both of the University of California, Berkeley, had independently begun their own x-ray irradiation experiments in January 1927. They hoped that these efforts would generate new traits in *Nicotiana tabacum*, the cultivated tobacco plant, for use in genetic and evolutionary studies.

Goodspeed and Olson's x-ray investigations extended in new directions a research program on the biology of tobacco that had been ongoing at Berkeley for more than two decades. Around 1905, William Setchell, the chairman of the Department of Botany and botanist at the California Agricultural Experiment Station, began collecting species of the genus *Nicotiana*, the tobaccos, for display in the university's botanical garden. As Setchell sorted his initial specimens, he realized that the taxonomy of *Nicotiana* was in disarray. There were many distinct cultivated types, but their relationships to one another were uncertain and the nomenclature applied to them confused.[39] Beginning in 1909, he enlisted the aid of Goodspeed, then the new assistant in botany, to sort out the taxonomy of the genus. In 1912, Roy Clausen, a biochemist and plant pathologist, joined the tobacco study as well. Gradually, Goodspeed and Clausen took over the project entirely. In the ensuing years, the two turned from comparative hybrid analyses to Mendelian studies to cytogenetic analyses, keeping with the same goal of determining the evolutionary history of the genus.[40]

Although Goodspeed and Clausen located many existing genetic differences among the tobacco varieties they analyzed over the years, they rarely encountered significant mutations in their research. "Now and then" some heritable alteration was found among the many thousands of tobacco plants in the Berkeley collections. Goodspeed estimated that perhaps "in every two or three thousand plants" a solitary individual with a change in a lone trait was found.[41] For a biologist interested in heredity and evolution, this was a disappointment.

Goodspeed turned to radiation treatment in early 1927, apparently hoping to jolt the tobaccos out of this complacency.[42] He had been encouraged to pursue this avenue by Olson, a professor of chemistry who was then particu-

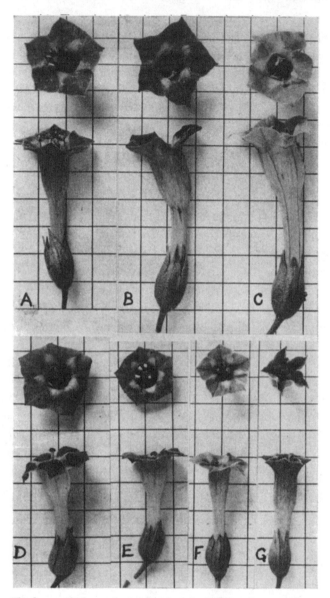

FIGURE 3. The flowers of tobacco plants grown from irradiated seeds as part of a 1927 experiment display significant variation. "A" is a flower from an untreated control plant. From Thomas H. Goodspeed, "Cytological and Other Features of Variant Plants Produced from X-Rayed Sex Cells of Nicotiana Tabacum," *Botanical Gazette* 87, no. 5 (1929): 570.

larly captivated by the study of radiation. Olson had conducted experiments to chart the effects of x-rays on chemical reactions, measured the extent to which x-rays were able to penetrate different elements, and, just a year earlier, collaborated with a bacteriologist in an investigation of the effects of radiation on bacteria and viruses.[43] It was the latter research that had led him to propose a similar project to Goodspeed.[44] Although there is no record of the outcomes they anticipated at the outset, Goodspeed and Olson later described their joint effort as one intended "to produce heritable variations" in *Nicotiana*.[45]

In the first stage of the experiment, conducted in January 1927, Goodspeed carried two *Nicotiana tabacum* plants from his greenhouse collection to a fan-cooled lead room in the chemistry building. He removed all the flowers and fully developed seeds from each, leaving only the unopened buds. By trimming and arranging the branches, Goodspeed managed to situate the remaining buds so that they would all be approximately the same distance from the radiation source, a Coolidge x-ray tube. Then, with Olson's assistance, he made the exposures. They irradiated one plant for ten minutes and the second for twenty. When the treatments were finished, Goodspeed took the plants back to the greenhouse. After a couple of days, the smaller of the buds on his tobacco plants fell off, evidently destroyed by the radiation treatment. More of the larger buds survived, however, and Goodspeed allowed these to flower, self-pollinate, and set seed as usual. Over the summer, he grew more than one thousand tobacco plants from these seeds.[46]

It was among these first-generation plants, grown from the seeds of the irradiated tobacco plants, that the more interesting effects of the radiation exposure became apparent. Many featured novel characteristics. In fact, the quantity of unusual plants that sprouted that season astonished Goodspeed. For years, he had been hard-pressed to locate individuals with new traits or unusual chromosome arrangements. X-rays, by comparison, appeared to produce these in abundance (fig. 3). He and Olson reported finding among the progeny of the x-rayed plants a wide range of morphological variations in which "no two were identical." There were dwarf plants and tall plants; there were plants with large, small, smooth, wavy and corrugated leaves; the leaves were in some cases dark green and in others gray green; the flowers turned out "very small, small, long, fluted, . . . folded, notched, ten-sided, light pink, pink, lively red, reddish purple," and so on. The incredible plasticity of the organism under the radiation treatment, especially in light of its prior immutability, seemed to be an immediate confirmation of Goodspeed and Olson's initial goal. "It would appear that we now have a method of producing at will an extensive series of variations in *N. tabacum*," they wrote later that year. All one needed was an x-ray tube.[47]

Speeding Up Evolution

Before Stadler or Goodspeed and Olson could finalize their research, radiation-induced mutation was an international news story. In July 1927, the *Drosophila* geneticist Hermann Muller, of the University of Texas at Austin, announced in the journal *Science* that he had caused what he called the "artificial transmutation of the gene." He described how flies exposed to massive doses of x-ray radiation produced offspring displaying many apparent genetic changes. "Several hundred mutants have been obtained in this way in a short time," Muller claimed, "and considerably more than a hundred of the mutant genes have been followed through three, four or more generations."[1] His publication prompted Stadler to announce the results of his research on x-ray-induced mutation in barley and maize, and Goodspeed and Olson to do the same for their study of tobacco.

Dizzying accounts of scientists' success in "speeding up evolution" followed almost immediately in American newspapers and magazines. Muller and Goodspeed encouraged early and euphoric reports, while Stadler remained more reserved. It proved difficult, however, to be skeptical in an atmosphere of eager interest and abundant hopes. Although news reports drew on previous discussions of the potential value of induced mutation, such as those inspired by the research of Hugo de Vries and Daniel MacDougal, the celebratory coverage of x-ray research in American newspapers and magazines in the late 1920s and 1930s went far beyond earlier accounts in characterizing this research as revolutionary. Biologists and breeders were cast as having achieved unprecedented control over the form and evolution of living things, control that would soon produce dramatic changes in agricultural production and might well also address concerns about eugenic improvement and spur industrial progress. In light of such expectations, the research into potential agricultural uses of x-ray-induced mutation in the 1920s and 1930s seems almost modest. The longstanding expectations associated with the ability to induce mutation—an ability once only imagined and now seemingly reality—were intensified by the still more credulous reporting of 1927 and afterward. The result was a context in which in this area of research could hardly be neglected.

Those working with *Drosophila* had been perhaps the most active on the question of how to provoke changes in genes and chromosomes, as evidenced in their studies of temperature, radiation, and other treatments on linkage and crossing over—and especially in their research on mutation. In the fly lab at Columbia and elsewhere, researchers by 1925 had tested radiation, chemicals, and changes in the physical environment of their flies, all in hopes of generating the genetic mutations so useful in their experimental investigations. They raised flies under varying conditions, including moist and dry, hot and cold, and with light and without light. They studied the effects of crossing individuals from widely distant places. Thomas Morgan himself varied "temperature, salts, sugars, acids, alkalis," but he concluded that none led to mutation. Some of these experiments had appeared to produce mutations, in particular those with radium, but these were notoriously difficult to confirm. After all, spontaneous mutations occurred without such stimuli. Who was to say that the treatment had caused the change?[2]

One *Drosophila* geneticist who had been especially eager to find a means to produce or control mutation—and to provide definite proof of having done so—was Muller. As an undergraduate at Columbia University, Muller had participated in the early studies of *Drosophila* in Morgan's laboratory. Although something of outsider in the fly group, and often at odds with its other members, Muller nonetheless counted among its key contributors between 1912 and 1915. As a graduate student, he pursued lines of work that proved integral to the development of *Drosophila* genetics and to the chromosome theory of inheritance.[3]

For much of his early career, Muller focused on mutation, driven in this research by his desire to use the knowledge of genetics and heredity to place evolution under more direct human control. As he described in 1916, "The central problem of biological evolution is the nature of *mutation*, but hitherto the occurrence of this has been wholly refractory and impossible to influence by artificial means, tho a control of it might obviously place the process of evolution in our hands."[4] Muller's first project in this area, carried out almost entirely by his friend and fellow geneticist Edgar Altenburg, attempted to establish a baseline rate of mutation in *Drosophila*. Working on the hypothesis that recessive lethal mutations were likely the most common mutations, Altenburg screened flies specifically for these. He found an astonishing number during his first trial in 1918, indicating such a high rate of mutation that he initially declined to publish the results.[5] This collaborative work on the rate of mutation stemmed directly from Muller's interest in finding a means of altering it. He felt it necessary to establish a baseline rate so as to know conclusively whether some other treatment, such as temperature change or exposure to radium, altered this rate—as geneticists had hoped and speculated for nearly two decades. Muller was soon pursuing these other experiments, too.

In November 1926, Muller, now working at the University of Texas at Austin, attempted x-ray irradiation of flies with the aim of providing conclusive evidence that radiation could alter the rate of mutation. His research up to that point, along with that of many other geneticists who had apparently produced mutations through various methods in the course of their research, had led him to believe that such an effect was likely.[6] With the assistance of a local physician, he treated flies with radiation for periods of time ranging from twelve to forty-eight minutes. Then he mated the flies with a specially bred stock so that any recessive lethal mutations created by the treatment would be revealed in their offspring.[7] When Muller began to examine the second-generation flies, it was immediately clear to him that the radiation exposure had altered the rate of mutation and to an unprecedented degree. He reportedly did not leave the laboratory that first night in his eagerness to record each new mutant he discovered.

Muller kept up the trials and his analysis through the spring of 1927, and by summer, he had prepared a brief note on his findings, "Artificial Transmutation of the Gene," which he sent to *Science*.[8] The rates of mutation that Muller claimed to have obtained through the x-ray method were spectacular: a rise in the mutation rate of some 15,000 percent in germ cells that had received the heaviest radiation treatment. He had focused, as in his earlier research, on uncovering recessive lethal mutations. These lethals were usually "invisible" to geneticists, who tended to focus on changes in characteristics such as eye color or wing shape, and not on those mutations that simply led to a nonviable offspring (hence their invisibility). Muller's method allowed the invisible lethals to be quickly counted—and they occurred in abundance among the offspring of the x-rayed flies. Muller had also found visible phenotypic changes in his treated cultures. This finding likely bolstered his confidence, and that of his colleagues, that the x-ray treatment was producing mutations equivalent to those that were already known.[9] With detailed data from thousands of flies, both treated and control, and a straightforward system to screen for genetic changes, Muller appeared to have produced irrefutable evidence that humans could affect the mutability of the gene.[10]

Muller understood that the genetic alterations he had produced in his fly cultures included both changes in chromosomes such as translocations and inversions and what he called "truly mutational" events.[11] By the latter, Muller meant a heritable effect not due to some gross change in the shape or arrangement of the chromosomes, or the destruction of the genetic material, but due instead to an alteration in an individual gene visible as a change in a Mendelian unit character. He presumed that these "true" mutations were analogous to the types of genic changes underlying the natural process of evolution.

This interpretation of his results encouraged Muller to speculate further about his long-held ambition of using genetic research to gain control over the evolution of species. As he suggested in 1927, as a result of his findings, biologists could now confidently take advantage of x-rays to create "in their chosen organisms a series of artificial races" to use in genetic analyses. They would no longer have to wait for an experimental organism to mutate spontaneously in order to pursue genetic research: "it should be possible to produce, 'to order,' enough mutations to furnish respectable genetic maps." By this Muller meant not that biologists would be able to produce specific mutations, rather that they would be able to produce as many random mutations as they wanted, when they wanted—and that achievement alone would be a boon to research. Muller also indicated the potential for future agricultural applications. He imagined the frustration of breeders, constrained to "remain content with the mere making of recombinations of the material already at hand, providentially supplemented, on rare and isolated occasions, by an unexpected mutational windfall." By comparison, his method would make it possible to produce "to order" these much-desired mutations, and therefore might prove as useful for the practical breeder in developing new crops and animals as for the geneticist studying mutations in the laboratory.[12]

Muller's publication of these findings pushed Stadler and Goodspeed to come forward with their related research; however, each was likely already aware that others were in hot pursuit of the same goal. A few months earlier, an article in the *Proceedings of the National Academy of Sciences* had detailed experiments on the effects of radiation on heredity undertaken by C. Stuart Gager of the New York Botanical Garden and the biologist Albert Blakeslee of the Carnegie Institution of Washington's Department of Genetics (previously, the Station for Experimental Evolution) at Cold Spring Harbor. Inspired by de Vries's mutation theory and motivated by MacDougal's and Gager's earlier research on the effect of radium emanations on plants, Gager and Blakeslee had exposed jimson weed, *Datura stramonium*, to radium.[13] Although the two had debated for a number of years whether the mutations apparently produced by exposure to radium could be attributed with certainty to the treatment, by the mid-1920s they were convinced. Their 1927 article announced that radium had caused the alteration of both chromosomes and genes.[14]

X-rays, however, made the big headlines. The December 1927 meeting of the American Association for the Advancement of Science (AAAS), the country's largest general scientific society, saw Stadler, Goodspeed and Olson, and Muller present their x-ray mutation research.[15] Their findings inspired genuine excitement among attendees. A special committee presented the annual American Association for the Advancement of Science Prize, a $1,000 award for a "notable contribution to the advancement of science," to Muller for his investigations.[16] The zoologist Winterton Curtis subsequently conveyed biologists' enthusiasm for this research in a public lecture via radio broadcast entitled "New Miracles with X-Rays."[17] According to Curtis, irradiation was "a new technique before which old problems may fall." It could selectively destroy cells "as though by a surgical operation of surpassing delicacy," change genes and patterns of inheritance, and expose physiological processes to new examination. To the well-known medical applications of radiation such as in the treatment of cancer, scientists would add novel uses of the rays in agriculture. In his estimation, no major subfield of biological investigation would escape the effects of exposure to x-ray technologies.[18] Journalists in attendance got caught up in these speculations, too. A reporter for the *New York Times* came away from the meeting amazed by biologists' newly acquired control over plants and animals. He believed Stadler to have demonstrated that x-rays could be used to create "entirely new species of grain." "Here we may have a new method for Burbanking flowers and plants for man's benefit," the reporter declared, comparing the young Missouri geneticist to the world-famous horticulturist Luther Burbank.[19]

Stadler, Goodspeed, and Muller took three different pathways to their discovery of the mutagenic effects of x-rays. Stadler stumbled onto obvious evidence of mutagenesis while using x-ray radiation for a set of genetics experiments aimed at altering expected patterns of heredity via a different route. Goodspeed, searching for new types that would be helpful for his research program in evolutionary biology, was led by Olson to believe x-rays would aid in this endeavor. Muller, long fascinated with the prospect of controlling evolution through control of mutation, first developed an experimental system that would enable him to reveal the occurrence of mutations more effectively; only later did he apply x-rays in hopes of doing so. Each knew the importance of his accomplishment. Never before had there been a widely confirmed means of producing genetic mutations on demand.

Of the three, only Stadler harbored doubts as to whether x-rays and radium induced "true" gene mutations—that is, not a change in chromosome morphology but the conversion of a gene to a new allelic form. He offered a number of reasons for these doubts. For one, cytological analyses (for which microscopic slides were prepared that revealed the cell nuclei, including the chromosomes) made visible many "chromosome aberrations" resulting from irradiation treatment. This suggested that observed mutant traits might be linked to these gross physical alterations of the chromosomes, not the hoped-for genic changes.[20] In Stadler's opinion, the linkage maps of maize were not yet sophisticated enough for him to decide whether certain observed changes in heritable traits were a result of gene mutation or chromosome aberration.[21] Furthermore, he was adamant that the recessive mutations that appeared to result from the treatment did not necessarily mean that the dominant gene had been converted into a recessive, though the behavior of the mutations in inheritance did suggest this. It could also indicate that the dominant gene was absent entirely, in other words, that the "mutation induced is simply the destruction of a gene."[22] Stadler proposed that this phenomenon accounted for many observed spontaneous mutations and nearly all radiation-induced mutation.[23] Muller, by comparison, was quite confident in 1927 that his work had produced "true" gene mutations and not just loss of genes, rearrangement, or damage to the chromosomes, even though the latter were also found among the irradiated flies.[24] Goodspeed agreed. He and Olson assumed that their work confirmed Muller's, even though subsequent cytological analyses also turned up visible alterations of chromosomes.[25]

The difference in the assessments made by these scientists manifested itself most sharply in the willingness of each to advocate his work as directly relevant to agricultural breeding. Muller clearly had this in mind from the outset, and his initial, somewhat speculative, vision for future agricultural uses of radiation articulated in *Science* became more concrete in subsequent accounts. In a 1928 interview, he described the urgency of extending induced-mutation research to domesticated animals and crop plants.[26] He believed that breeders needed to find new methods in order to truly master the development of crop plants or animal stocks. As he explained elsewhere, "We must, in a few years, be able to produce as many mutations in our animals and plants as ordinarily occur in a thousand years, if we would succeed in properly subjugating the genes of our crops and stocks." Achieving this control was crucial, especially for economic growth and social stability: "The fortunes of our state will rise and fall largely as the quality and quantity of our cotton, our corn, our beef, and our other animal and plant support, wax and wane."[27]

Stadler, by comparison, understood the application of x-rays to breeding to be "much less sensational than is sometimes claimed" and in fact often despaired of the bold assertions of Muller and others.[28] There were a number of reasons for Stadler's wariness, which went beyond his skepticism about whether the induced changes were in fact genetic mutations. He pointed out that "a great wealth of germinal variation," the accumulation of all evolutionary history, already existed in plant species and could be brought into established varieties as needed by hybridization.[29] Unlike Muller, he was familiar with the diversity existing in field crops under cultivation. He was also skilled in creating varieties through the combination of known types, and thought that these methods would not immediately be surpassed by the x-ray production of genetic diversity. What's more, his experiments had given no indication that induced changes were favorable. In fact, there was reason to believe, as Stadler did, that they were almost all deleterious. Even if a favorable mutation were to be induced, it would no doubt be accompanied by damage elsewhere in the chromosomes simply because organisms had to be completely bombarded with radiation in order to turn up altered characters in any meaningful numbers. The only areas in which Stadler felt breeding via induced mutation might be useful were in introducing variation into highly inbred lines, such as those used in hybrid corn, and in producing new somatic mutations in fruit. These were areas of breeding where techniques of introducing genetic variation through hybridization were cumbersome or did not apply.[30]

Goodspeed saw more promise than did his fellow plant biologist Stadler. Like Muller, he was stunned by the abundance of changes that turned up after irradiation. These changes were not simple chlorophyll changes such as those Stadler had produced in his barley plants, but wide-ranging effects in color, flower and leaf morphology, and plant architecture. And they had occurred in an organism that Goodspeed had understood to be stable. Influenced no doubt by the extraordinary responsiveness of his tobacco plants to x-ray exposure, Goodspeed felt certain of the relevance of his research to agriculture. As he described at a public lecture in Berkeley in 1929, the ability to bring about mutations through exposure to x-rays or radium suggested "that new varieties of importance in agriculture may be produced with a much greater rapidity than has ever before been possible."[31]

Muller and Goodspeed's optimism, shared by many others, left Stadler somewhat isolated. In the late 1920s, his skepticism was lost in the fervor that surrounded induced-mutation research and the new opportunities created by the artificial modification of genes. Just as the initially divergent research interests of Stadler, Goodspeed, and Muller eventually converged on a single end—finding a method that would induce heritable changes—so, too, did the imagined consequences of their research narrow to a single narrative once it hit newspapers. Reports equated the ability to induce mutation on demand with a new capacity to direct evolution and emphasized the potential practical payoffs of having such a capacity. This was, in some respects, a well-established idea. The science of genetics had from its earliest days inspired claims to having enhanced human control over the direction of evolutionary change, hence the enthusiasm of farmers, breeders, and eugenicists for this developing discipline at the turn of the century. In 1927, however, the x-ray mutation work introduced a new notion of how genetics research might be applied: it now offered the ability not only to guide more precisely the direction of evolution but also to accelerate evolution itself. That was a striking new capacity for human beings—and one that journalists made much of.

National news coverage of x-ray-induced mutation took varied forms, nearly all of it celebratory. A 1930 Associated Press report on Stadler's research summed things up in what by then had become a typical manner: "New plant varieties, improved animal breeds and hitherto undiscovered evidence on the mechanism of heredity, variation and evolution are expected from the X-ray laboratory."[32] In many accounts, x-ray-improved varieties appeared to be extremely easy to produce. As a 1928 Science Service survey emphasized, x-ray mutation research "hold[s] out the possibility of developing new varieties for the farmer and gardener and rancher with a rapidity that would have been dismissed a year or so ago as fantastic and visionary."[33] According to another newspaper report, Goodspeed and Olson had produced over two hundred new varieties simply by applying x-rays to commercial tobacco. The article also described how the experiments of the Berkeley duo had produced a supertall plant, twice the height of normal tobacco, with more leaves and greater vigor, an outcome that surely confirmed both the speed and immense rewards of the x-ray method of breeding (fig. 4).[34]

At the heart of this faith in faster production was the notion that so-called artificial mutations, created through x-ray exposure, granted humans control over the speed of evolution. Muller cast his own research along these lines, speaking of creating mutations on demand as a way to "defeat time," and news reports similarly tended to interpret a mutation rate 150 times faster than expected as, in the words of one United Press report, "condens[ing] into one year all the changes that in the past have required 150 years to develop."[35] The very idea of accelerating evolution through induced mutation reflected contemporary understandings of evolution in which genetic mutations were a crucial source of the variation in populations on which natural selection acted. But "never previously," the typical story went, had it "been found possible to make them [mutations] occur oftener."[36] Scientists had beat nature at its own game. One benefit of this achievement was that scientific research could progress faster. Whereas before the discovery "plant variation experiments took years," with botanists waiting for "nature to work the changes," x-rays would

FIGURE 4. A "towering new tobacco plant" dwarfs its purported creators Thomas Goodspeed (*left*) and Axel Olson. A more typical specimen of cultivated tobacco grows between the two men. From *Popular Science Monthly*, March 1928, 51.

now speed them along.[37] Another payoff, of far greater interest in the press, was the faster production of new crop varieties. "Assisting or forcing nature in some way, so that new things will be produced faster than at the old, poky rate, has for centuries been the breeders' dream," explained one science writer in predicting the future marvels to result from induced-mutation research.[38]

Although most reports about accelerating evolution dealt with studies of plants and *Drosophila*, the radiation research most commonly pursued within the scientific community, some also covered the extension of x-ray mutation

FIGURE 5. A 1931 news report details the "miracles" expected from biologists working with x-rays. The illustrations show a flowering plant under an x-ray device, an irradiated chicken that reportedly developed more quickly, and "Jack's beanstalk of gigantic proportions," which "may become a reality under X-ray treatment of the seed." From Eleanor Early, "X-Rays to Produce Giant Flowers, Fruits, and Animals," *Public Ledger Sunday Magazine*, 26 April 1931, 6.

to the improvement of animals (fig. 5). William Dieffenbach, a physician at the New York Homeopathic Medical College and Flower Hospital, drew attention for his application of x-rays to chicken eggs in a series of experiments he began in 1928. The experimental effort proved that Muller's findings held for higher-order animals, or so Dieffenbach thought, and also that useful improvements in these animals could be made using x-rays.[39] Others merely speculated. Leo Brosemer, a livestock breeder and former USDA employee, claimed to have analyzed the relevant x-ray research and determined that it would lead to faster racehorses, more productive milk cows, bigger pigs, and "super hens." He emphasized the speed and efficiency of the method, describing how alterations that usually took one hundred years would take a mere three or four.

This in turn would have effects that could alter far more than the shape, size, and speed of animals. "If breeders can . . . manufacture farm animal types to order and in numbers, they will revolutionize the diet, the customs, and the mode of living of the civilized globe almost as completely as the invention of the steam engine," he predicted.[40]

Brosemer's grand vision for an x-ray-improved society was modest in comparison with the route envisioned by a few others, who saw mastery of x-ray evolution as a means eventually to control human evolution. Tantalizing, or perhaps terrifying, his readers with visions of a world populated by one-thousand-year-old highly talented men and women in perfect health, the reporter George Gray described how such a thing might be accomplished if the "basic problem" of evolution's slow pace could be solved. That scientists were on the way to accomplishing such a thing was evident from the x-ray research of Goodspeed, Stadler, and others, according to Gray, and the remarkable changes they had effected in a short time.[41] If plant evolution could be manipulated, surely human evolution could be, too. The journalist Eleanor Early, evidently thinking along similar lines, wondered, "If an X-rayed plant produces strange seedlings, how about an X-rayed hen hatching a new sort of chicken? How about X-rays producing a giant superman?"[42]

The idea of speeding up evolution had appeared in other American conversations in the preceding three decades, most often in reference to social change and not biological evolution.[43] The exception to this rule was in agriculture, where prior to 1927 breeders were sometimes declared to have accelerated the evolution of varieties through selection and hybridization. Burbank was the exemplar. "Darwinism taught us that species arose only through slow ages of change by the gradual process of natural selection accumulating its effects for thousands and even millions of years; but Luther Burbank shows that *man* can produce species and do it in a dozen summers!" exclaimed one journalist in 1905.[44] After the demonstration of x-ray-induced mutation, Burbank's methods were no longer depicted as revolutionary, but rather as the quintessence of slow and old-fashioned methods. According to lore, Burbank had planted one thousand seeds of each plant he hoped to improve, and only one in that number would have a new trait helpful to his breeding projects. "You can imagine how much work that was for Mr. Burbank and how many miles of land he needed to grow his experimental plants," one reporter declared.[45] In the era of the x-ray breeder, all this would change. Breeders now had the option of causing mutations to appear "at will." No longer did they need thousands of plants and miles of productive land. "If Burbank had known what we know now, he could have done his work on a city lot," claimed a physician who had recently begun his own experiments with x-rays.[46]

One of the imagined benefits of this acceleration, besides economy of space and speedier production, was that it would allow scientists and breeders to "make new plants to order." X-ray breeding was likened to the streamlined processes of the factory floor or the blueprint planning of an architect, implying that a previously slow and haphazard process and been rendered compatible with the machine age. Just as the animal breeder Brosemer had described the "manufacture" of "farm animal types to order and in numbers," other reports summarized x-ray research as showing, as one headline declared in short, "New Life Made to Order."[47] Suddenly, it seemed possible to go beyond building a crop plant from component parts, constructing and assembling from what nature had provided, and instead to fabricate the very traits from which a plant would be made. Radiation would generate them de novo. *Popular Mechanics* likened the process to one of machine assembly: "As shifting of wheels, nuts, screws and bolts changes a machine, so the X-ray changes the life form and there is a new creation."[48] In making a connection between mechanical and genetic components, journalists followed the lead and language of the scientists on whom they reported. In 1927, Muller made a similar comparison between dead machines and living organisms: "If we are to . . . make continued rapid progress in the shapes and characters of our living possessions, comparable with what we are now doing with inanimate objects in our mechanical arts, we must make use of some new tools such as the x-rays, which will keep the ball of change rolling."[49] Others linked mutations and machine parts still more directly. "We shall be able to manufacture a series of artificial strains. It may soon be possible to order sufficient mutations at will," declared the British geneticist Francis Crew.[50]

It was no small ambition to hope that breeding might in time more closely resemble manufacturing, especially in the United States in the 1920s. The decades around the turn of the century had witnessed the rapid development of American manufacturing capabilities, placing the country at the forefront of global industry. Mass production in particular captured attention as a revolutionary approach to the creation of a range of consumer goods.[51] The icon of such production in the 1910s and 1920s was undoubtedly the Ford Motor Company, whose adoption of moving assembly lines along with other machine and labor innovations had led to dramatic increases in productivity and to Ford capturing up to 50 percent of the automobile market in the United States by 1921. It also led to the hope that other areas of production might be made more productive and profitable through similar changes.[52] These areas of production included agriculture. Enterprising reformers thought that the typical American farm could be transformed into a large-scale enterprise, centered on the production of a specific crop or crops, that utilized new ma-

chine technologies and a routinized workforce to achieve higher through-
put and greater efficiency.[53] These hopes for the improvement of agriculture
through greater alignment with industrial principles gained still greater urgency during the 1930s. By that time, many farmers had been struggling for
more than a decade, the economic woes of the Great Depression intensifying
the effects of an agricultural depression that had begun in the early 1920s.[54]

With this industrial and agricultural context in mind, it is easy to understand why many Americans considered "speeding up evolution" such a
promising advance. Making evolution run faster did not mean that it would
accelerate beyond human control. On the contrary, it would bring the creep-
ing natural pace of this process into sync with the more demanding tempo of
modern American society. It seemed likely to allow plant breeders the option
of creating varieties more efficiently and more rationally. At a time when farmers, extension agents, bankers, and government officials alike were increasingly concerned with streamlining, mechanizing, and otherwise industrializing agriculture to bring it in line with the ideals of modern production in the
United States, this was just another promising route to making farm output
more efficient.[55] The "long, uncertain period of leisurely-occurring" mutations, the "old, poky rate" of change, the "slow crawl" of evolution—the limitations placed by nature on agricultural and horticultural production—would
be eliminated.[56] In many ways, Winterton Curtis's 1927 depiction of x-rays as
a source of "miracles" captured exactly the prevailing sentiment in which the
power to directly manipulate genes to the advantage of humans was in hand
and would bring with it sweeping social change. Amid this swirling discourse
of newfound power and profound change, many tried to turn such claims into
something more than sensational speculation.

X-rays in the Lab and Field

The first step in transforming speculation into reality was to step up research. Shortly after the Nashville AAAS meeting, Winterton Curtis and a few others conspired to establish a fund within the National Research Council (NRC) for the support of further investigations into the biological effects of radiation. The Committee on Effects of Radiation upon Living Organisms (hereafter, Committee on Effects of Radiation), which was convened from 1928 to 1934, supported research on the nature of induced mutation in plants and animals, as well as a range of radiation-related subjects including the effects of radiation on growing tissues, embryonic development, whole organisms, and organic substances such as proteins and vitamins. According to its statement of purpose, the committee intended to support "pure science" and produce information that addressed "fundamental problems" in the fields of morphology, heredity, and physiology—but it was also interested in potential applications of this research, particularly in medicine and agriculture. By 1934, the committee, supported by both private philanthropies and industrial producers, had distributed research equipment and funding to more than forty American researchers whose work related to the biological effects of radiation.[1]

As a result of such attention, induced mutation quickly became both the subject of further biological investigation and an accepted experimental tool, nowhere more so than in genetics. Researchers who chose to pursue such work benefited from the growing interest and speculation about induced mutation in the form of greater institutional and financial support. Industrial firms in particular proved eager to come to the aid of the Committee on Effects of Radiation, donating funds and equipment and offering technical assistance. Universities and departments, sensing this was a cutting-edge and media-friendly topic, ratcheted up their support of x-ray research. And plant breeders, whether employees of state agricultural experiment stations or commercial agricultural and horticultural producers, sought that most promised payoff of induced-mutation research, improved crops. One of the first questions that the geneticists interested in radiation effects had to address was whether the mutations apparently produced by radiation treatment were truly equivalent to those that appeared spontaneously. Stadler objected that, in maize at least, cytological studies showed that the observed effects resulted not from genic changes but gross changes in chromosome morphology. But he (and others) had additional concerns. For example, if "artificial" mutations were the same as those that underpinned evolutionary change, surely some would have to confer obvious advantages, such that they might be selected for in nature, and yet exposure to x-rays seemed to produce mostly deleterious changes. Proving that radiation produced advantageous, or "progressive," mutations was also crucial to the argument that these would be useful in plant breeding.

Not everyone was interested in waiting for such evidence, however. Many simply forged ahead in the quest to create x-ray-improved varieties, irradiating anything from apples to cotton to gladiolus flowers. The need for new innovations, for novel traits and therefore novel types that would either meet longstanding demands or generate new ones, proved incentive enough to begin experiments.

In June 1928, as the Committee on Effects of Radiation was just getting organized, the chairman of the NRC's Division of Biology and Agriculture, William Crocker, wrote strongly in favor of its formation: "I do not believe the importance of radiation effects in biology can be over-emphasized." He mentioned a number of biologists and their research findings, whose work he took as proof that "irradiation gives a new instrument and technique for controlling and modifying inheritance, and development of organisms."[2] Curtis went still further, especially when soliciting funds, referring to x-ray treatment as "a procedure infinitely more delicate than anything previously available" and "the real beginning of a physico-chemical analysis of life processes."[3] The hopes pinned to x-ray experimentation were ambitious, and the committee vigorously sought evidence of bringing these to fruition in the years that followed, primarily by supporting biologists already working in this area.

Before the Committee on Effects of Radiation could disburse funds, however, it had to raise them. Its principal sources of funding, as was the case in many areas of research in the natural sciences in the 1920s and 1930s, were private philanthropies.[4] Two such institutions, the Commonwealth Fund and General Education Board, together kicked in $125,000 to support the committee over a five-year period. The committee initially hoped to obtain assistance in the form of cash sponsorship from industrial producers as well, especially manufacturers of x-ray equipment. Its organizers no doubt assumed that these producers would be happy to support research that promised to increase the value of their commercial goods—and they were right. In response to a solicitation for funds, the Victor X-Ray Corporation (by that time, a GE-owned company) offered a cash contribution and loan or gift of up to $10,000 worth of equipment. As the president of the corporation explained, he and his colleagues were "heartily in sympathy with research work along this line." After all, the research would "mean a very definite step towards standardization" in therapeutic uses of x-ray equipment, which would "in turn . . . place the manufacture and sale of physical therapy [x-ray] equipment on a much more satisfactory basis."[5]

Even in the midst of the greatest economic crisis the United States had yet seen, around twenty companies—mostly producers of x-ray equipment, ultraviolet lamps, and other radiation-related medical devices—came forward to support the study of "fundamental problems" in heredity, evolution, and physiology. Only a handful provided cash donations, and none came close to matching Victor X-Ray's eventual $5,000 contribution, but they proved happy to arrange for the loan or donation of equipment essential to the continuing radiation investigations. Members of the committee speculated that the economic depression was in fact aiding their efforts, with manufacturers less burdened by the "problems of an expanding production" and therefore "more willing" to support research.[6] There might have been a kernel of truth to this speculation, if the willingness of firms to also donate the time and energies of their staff was any indication. Willis Whitney of the GE research laboratory, already a champion of investigation into radiation effects, was wary of sending resources to fund work that might duplicate what was going on in his own lab (though the GE corporation and some of its subsidiaries did donate cash and equipment). He nonetheless offered up the advice and assistance of his employees whenever the committee's grantees might need it.[7] The Bausch and Lomb Optical Company, Eastman Kodak Company, and Westinghouse Manufacturing and Electric Company similarly offered support in the form of advice and apparatus.[8] Other companies might have seen the work of the Committee on Effects of Radiation as a way to have useful research conducted on the cheap, without the use of their own staff. The Wappler Electric Company was glad to donate all manner of devices and components such as transformers, x-ray tubes, tube holders and stands, and so on. It also offered to share anything relevant that its own researchers learned, on the understanding that "in return, we would appreciate to learn of the results of these [the committee-sponsored] investigations."[9]

The money, materials, and advice made available by these firms, and by the Commonwealth Fund and General Education Board, made their way to a number of researchers across the country. The committee's grantees included those already entrenched in radiation studies, whether of genetic or other physiological and developmental effects, alongside other researchers who now saw radiation as relevant to their research. Their objects of study included fruit flies and maize, wasps and tomatoes, rats and bivalves, and several species of fish and types of microorganisms, and they utilized various forms of radiation including x-rays, radium emanations, ultraviolet radiation, infrared radiation, solar radiation, and so-called mitogenic rays, radiation thought to be produced by the cell itself during cell division. Their investigations sought the effects of radiation on enzymes, cells, pollen, nutrient uptake, metabolism, development—and, of course, hereditary effects in genes and chromosomes. Within the last category, Goodspeed, Muller, and Stadler numbered among the first beneficiaries of the program, but their work on induced mutation was not the only such research to benefit from the efforts of the committee. For example, Ernest Babcock, also of the University of California, Berkeley, received support for his efforts to induce mutations in the flowering plant *Crepis*, Frank Hanson and C. P. Oliver of Washington University for their work on *Drosophila*, Mary Stuart MacDougall of Agnes Scott College for studies of induced mutation in microorganisms, and E. W. Lindstrom of Iowa State for his efforts with tomato plants.[10]

As research into induced mutation expanded, a pattern established itself in which the departments where these studies had initially taken place became hubs for continued research. At the University of Texas at Austin, Muller's home institution, mutation studies in *Drosophila* proliferated. A handful of geneticists working alongside Muller searched for the mutagenic effects of different agents, from chemicals to temperature extremes to ultraviolet radiation. They also attempted to sort out the many distinct effects of x-ray radiation such as translocations, inversions, and mosaics.[11] Meanwhile, at the University of Missouri, Stadler and those who worked with him similarly undertook radiation studies with various aims in mind. In one study, Stadler compared the effects of radiation in polyploid crops (that is, those with more than two sets of homologous chromosome pairs), speculating that the duplication of chromosomes would reduce the observed frequency of mutation. He also began to investigate the effects of ultraviolet radiation. As his reputation and his research program grew, Stadler and those who came to study genetics with him at the University of Missouri continued to focus on mutation as a central concern of biological research and radiation as a key tool with which to conduct that research.[12]

X-RAYS IN APPLE BREEDING

FIGURE 6. A researcher demonstrates the use of a portable x-ray machine to irradiate apple trees in the hope of producing a somatic mutation. The apparatus is oriented so that the rays are concentrated on a single developing bud. From *Journal of Heredity* 21, no. 1 (1930), frontispiece, by permission of Oxford University Press.

For a brief period of time, Stadler also undertook breeding projects involving radiation. He collaborated with two USDA corn breeders, x-raying highly prized strains of inbred corn, "in the hope of inducing some variation that might permit further improvement by selection." He was excited by initial evidence that new variations had appeared, including plants resistant to common diseases.[13] Together with the horticulturist A. E. Murneek, he investigated the possibility of inducing bud variations in Golden Delicious and Yellow Delicious apples through the use of x-rays (fig. 6).[14] In 1932, an Australian breeder engaged Stadler to x-ray wheat in order to induce rust resistance, an effort that Stadler defended on the theoretical grounds that there appeared to be a single dominant trait for rust susceptibility in the variety and therefore an "outside chance" that it could be eliminated by radiation treatment.[15] None of these projects appear to have had any success, an outcome that likely reinforced Stadler's emphasis on x-rays and radiation as useless for plant breeding.

Radiation-induced mutations were also a favored research topic at the University of California, Berkeley. Inspired by the work of Goodspeed and Olson, Berkeley researchers undertook a diverse set of x-ray activities. One of the earliest manifestations of the growing interest in radiation was the installation of x-ray equipment at the experiment station.[16] By July 1929, members of the division reported having made 233 x-ray treatments of plants and animals. *Nicotiana, Crepis* (popular among the Berkeley geneticists as an experimental model), and *Drosophila* featured most prominently among irradiations made for "research purposes," but the staff irradiated almost anything of interest. One tally included "mosquito larvae, avocado seedlings, wheat, barley, and oat seeds, Pinus seeds and pollen, and certain algae and fungi."[17] These organisms represented the varied concerns of those at the station, from improving cereal crops and horticultural varieties, to studies of malaria mosquitoes, to research that could be used by the California timber and logging industries.[18] In addition to these various studies conducted by Berkeley staff, the Division of Genetics also provided irradiations "as a form of service," making its x-ray equipment and expertise available to researchers beyond the Berkeley campus.[19]

Of these many x-ray-related activities at Berkeley, Goodspeed's *Nicotiana* research was the most extensive.[20] In 1929, he was confident that the tobacco studies produced "qualitative alterations of the hereditary material" that could be produced "at will."[21] He stabilized several lines of the x-rayed tobacco by breeding them out for a few generations, and then shared these lines with other geneticists on request.[22] He was also eager to see whether tobacco farmers would be interested in the mutant lines. One variety carried a suite of desirable traits including more and bigger leaves. Goodspeed reported that it had been "examined by tobacco growers as an improvement on available commercial races."[23]

During the spring of 1929, Goodspeed collaborated with the biologist J. W. McKay in a study of x-ray effects in cotton. The effort further cemented his view that x-rays held considerable potential as a tool of plant breeding. McKay had recently come to work at Berkeley from the University of Texas where in 1927 Muller had encouraged him to attempt the irradiation of cotton, one of the state's key economic crops. Reportedly, he had hoped that such a study would confirm his discovery of induced mutation "in the case of a plant species of commercial importance." Following Muller's suggestion, McKay had applied x-ray treatment to the cotton plants and harvested the seeds. These he brought with him to his new stint at Berkeley, where he sowed them in the spring of 1929 and evaluated the results with Goodspeed. The most exciting outcome of the experiment—which resulted in many gross morphological changes—were a couple of plants that produced what McKay and Goodspeed described as "naked seeds." This meant that the seeds contained within the cotton boll floated free from the fibers, making the two easier to separate in ginning. This characteristic would mean more efficient processing of the fiber, assuming that it could be stabilized as a trait.[24] In other words, the cotton irradiation experiment seemed to have produced a clear demonstration of the potential of x-ray-induced mutation in plant breeding.

Not everyone at Berkeley was interested in introducing greater speed and efficiency into agricultural breeding, but even those who were not still saw x-rays as a way to speed along their research. The head of the division, Ernest Babcock, applied x-rays to his preferred experimental organisms, plants of the genus *Crepis*. By 1927, Babcock had been working with *Crepis* for over a decade in his studies of evolutionary biology.[25] Interested in the potential use of x-rays as a means to generate new traits for his studies in heredity and evolution, he attempted to use radiation to "test the possibility of producing at will new types of plants." Understanding evolutionary processes, he explained, required numerous plants of a given species with distinct variations, and "to wait for the chance appearance of these new types under natural conditions takes too long and progress is too slow." X-rays would accelerate the pace of his research into plant evolution by accelerating this basic biological phenomenon.[26]

For researchers at other institutions working on x-ray-induced mutation, the origins and interpretations of their research were inextricably linked to improvement. At the Texas Agricultural Experiment Station, Walter Horlacher and David Killough found evidence about the "universal application" of x-rays to the production of gene and chromosome changes to be so compelling as to demand the production of data "as rapidly as possible" on economic varieties. Horlacher, a professor of genetics at the Agricultural and Mechanical College of Texas, and Killough, the agronomist in charge of cotton breeding at the Texas station, irradiated seed of American upland cotton at the college hospital. The two found numerous unusual traits in the first generation grown from the x-rayed seed: dwarf plants, chimeric or variegated plants, and unusual leaf quantities and shapes.[27] Of all these developments, the one that proved to be of surpassing interest to Horlacher and Killough was an effect that x-raying produced in a particular strain of cotton, which was normally "virescent" or greenish-yellow in color rather than the expected deep green. Killough in earlier work had demonstrated that the virescent trait, a random mutation that had been found in a Texas cotton field in 1925, was a Mendelian recessive to the dominant green.[28] In plants of the virescent strain that Horlacher and Killough grew from x-rayed seed, however, new patterns of coloration emerged in which whole sections of leaves displayed a green color or no color at all. And in one "scraggly plant" of the virescent line, "a very poorly developed specimen growing prone on the ground," each and every leaf was entirely green as a typical cotton plant's leaves would be.[29]

It was in this unlikely specimen that Horlacher and Killough invested their hopes for having made a crucial contribution to genetics research as well as having demonstrated the usefulness of x-rays in agricultural improvement. They believed that the x-ray-generated effect of turning the virescent strain of cotton into a fully green plant demonstrated that mutation could be "reversed," meaning that the recessive trait apparently had reverted to the original dominant. It was, moreover, clearly a "progressive" change in which a sickly yellow tinge had been transformed to the lush green of healthy vegetation. Were this not enough, they identified a second induced progressive mutation among their plants: an alteration in leaf shape from "forked" to "normal." As Horlacher and Killough described, the alteration of leaf shape from forked to nonforked meant an increased leaf surface area, which in turn meant "more food-making surface."[30] They concluded that their research refuted the argument that "all of the induced changes are destructive, that no variation is produced by radiations that is of any value to the organism or to man." They hoped their evidence would spur breeders in x-ray work, for "it can leave little doubt but that favorable mutations can be induced in cotton."[31]

Still other agricultural station researchers deployed x-rays in the hope that these could someday—or immediately—be used in plant improvement. In 1936, the USDA *Yearbook of Agriculture* listed three other agricultural stations pursuing x-ray-induced mutation in cotton, although it mostly catalogued their failures: Arkansas ("no mutations were found in the plants developed from the treated seeds nor in the next generation"), California ("results so far very unsatisfactory"), and Louisiana ("several stunted and abnormal forms . . . but nothing of economic value").[32] Other articles in the 1936 *Yearbook* similarly marked the x-ray approach for special mention. It was featured among options offering "unusual possibilities" for breeding in the future and also described as a tool already deployed in station work.[33] The agronomist John Martin reported that sorghum breeders had corroborated their suspicions about unusual changes in milo, a small drought-resistant sorghum, by repeating the production of one such change with x-rays.[34] Breeding methods noted by tobacco experts included both attempts to double the chromosome complement with x-rays and the use of x-rays to induce mutations of single characters.[35] Efforts to tease out new traits in rice with x-rays were, by comparison, not a documented part of American breeding efforts, though the USDA agronomist Jenkin Jones described work by Japanese rice breeders to successfully induce mutations of all types, as well as triploid and tetraploid varieties, with x-rays.[36]

Commercial operations also supported x-ray breeding efforts.[37] One that quickly adopted x-ray methods was the Maui Agricultural Company, a producer of sugarcane. The company, in collaboration with the Hawaii Sugar Planters' Association, the Hawaiian Sugar and Commercial Company, and the Pioneer Mill Company, purchased x-ray equipment for its "factory laboratory" in January 1928, intending to explore the possibility of improving cane through induced mutation. J. P. Foster, the employee charged with leading the research, claimed that the company had "no expectation of immediately revolutionizing the sugar industry, or of producing marvelous new varieties of cane 'off the bat.'" But his intention was undoubtedly to produce an improved sugarcane by inducing mutations that would be useful to breeders. The industry reportedly had anticipated geneticists in this research program. The Hawaii Sugar Planters' Association claimed to have been interested in the possibility of inducing bud mutations through irradiation even before the announcements of Muller and others. According to Foster, "lacking the stimulus of the knowledge that success had been attained elsewhere," the company had not persisted as diligently as it should have in this effort. Now that recent scientific findings suggested a near-certain reward, it made the attempt in earnest.[38]

Where the Maui Agricultural Company struck out on its own, other commercial breeders relied on aid from established research institutions. In a widely reported case, W. Atlee Burpee & Co., a large mail-order farm-and-garden seed company, sent seeds to the genetics researchers at Berkeley in the 1930s to be exposed to x-ray radiation as part of the service irradiation program there. From a selection of garden seeds irradiated in 1933, a few promising variations were discovered among varieties of calendula, a colorful and easy-to-cultivate garden flower. Burpee breeders reselected and stabilized the plants displaying these variations over a number of years, and the company finally released them for sale in 1942 (fig. 7). Advertisements declared the "X-Ray Twins," also known as the varieties Glowing Gold and Orange Fluffy, to be "the first new flowers created by the use of x-ray."[39] According to a report about these calendulas in *Time*, they owed their existence foremost to Babcock at Berkeley. He was given credit for exposing the seeds to x-rays, a process that had "ionized—or 'electrified'—the seeds' nuclei, kneading their chromosomes into unusual patterns."[40]

Another commercial grower who sought assistance from the Berkeley irradiation program was the Czech-born horticulturist Frank Reinelt. He sent Babcock and his colleagues a stubborn delphinium hybrid with which he had struggled unsuccessfully to produce a red flower. He reportedly thought that exposure to x-rays might cause a "break" in which a new color, preferably red, would appear. Reinelt, known for his skilled improvement of specialty flowers such as begonias, camellias, and delphiniums, did not get the red strain for which he longed—but he did get several extremely large and densely flowered plants. According to the *Saturday Evening Post*, it was from this stock that

FIGURE 7. Burpee's "X-Ray Twins," a pair of calendula (marigold) flowers produced through exposure to x-ray radiation, were introduced in 1942 as the varieties Glowing Gold and Orange Fluffy. From *Burpee Seeds* [catalog] (1942), by permission of W. Atlee Burpee & Co.

Reinelt "evolved the Pacific strain," a delphinium that would grow as well in an outdoor garden as its more tender predecessors had in the greenhouse.[41]

These flower breeders may have indirectly benefited from the activities of the Committee on Effects of Radiation, which channeled money and tools from its donors to the Berkeley biologists in support of various x-ray investigations through the early 1930s. If so, then the flower breeders also indirectly benefited from the industrial interest in x-ray-induced mutation that the Committee on Effects of Radiation, through its fund-raising activities, both cultivated and cashed in on. But there was more than just a chain of grants that linked technological-industrial powerhouses like General Electric and family-run flower-and-vegetable seed concerns like W. Atlee Burpee & Co. Though their product lines looked wholly different, sprawling techno-industrial enterprises and flower-seed producers shared concerns: all had to generate, and continually update and improve, a line of desirable goods. For example, in order to stay ahead of its many competitors, GE required a continuous stream of innovations and improvements in the electrical devices it produced for industries and home consumers. W. Atlee Burpee & Co. similarly needed to provide more and better flower and vegetable varieties each season to satisfy its customers and keep on top of a competitive market. David Burpee, the head of the company from about 1915 until 1970, is well remembered as having taken an efficient industrialist's approach to turning out new varieties of popular garden flowers, dedicating significant resources to innovation and pursuing promising scientific and technological developments. The approach made his firm one of the country's most successful mail-order seed companies in the mid-twentieth century.[42] In other words, the demand for continuous innovation characterized both the production of electrical and mechanical goods and the production of biological ones. And, in the wake of the demonstration of x-ray-induced mutation, at least one company tried to bring these two areas of innovation and production together.

Industrial Evolution

Nowhere was the imperative to innovate, as shared across both mechanical and biological realms, more on display than in an experimental x-ray breeding program launched at the research laboratory of the GE Company in the early 1930s. While some biologists and agriculturists interested in the application of radiation in plant breeding tried to exploit the technique by bringing x-ray apparatus to their experimental fields, a pair of GE engineers brought the field to the x-ray laboratory to pursue the same interest (fig. 8).[1] In a small greenhouse erected on top of the main laboratory building at GE and in a garden patch out back, Caryl Haskins and Chester Moore investigated the effects of x-ray radiation on species ranging from gladiolus flowers to giant sequoias. This x-ray experimentation aimed both to demonstrate the usefulness of x-ray technology, an important area of innovation and production at GE, and to promote the expertise of a group of physical scientists and engineers in enhancing breeding practices. And, of course, it aimed to produce new plant varieties through irradiation.

The GE research laboratory, founded in 1900 as an extension of the company's central operations in Schenectady, New York, had been envisioned as a means to support the development of ideas and technologies that would keep its parent company atop the marketplace for electrical power and associated products in the United States and abroad.[2] Its early employees were, like those at other industrial laboratories in the United States and Europe, responsible for developing potential new commercial applications from existing knowledge—in other words, conducting focused development of future GE products or improvements on current products. GE took a comparatively broad view of the research relevant to corporate interests, so the laboratory was also a place for investigations into general scientific principles, without regard to their immediate ability to generate profit.[3] Nonetheless, it was seen

Science Links Industry and Botany

In This Outdoor Laboratory Scientists and Research Engineers Study the Results
Obtained from the Cultivation of X-rayed Seeds

FIGURE 8. "Science Links Industry and Botany"—or so the researchers involved with GE's project in induced mutation hoped. Here a GE employee tends to flowers grown from irradiated seeds in the garden plot at the GE research laboratory. From C. P. Haskins, "X-Ray and Cathode-Ray Tubes in Biological Service. Part II," *General Electric Review* 35, no. 9 (1932): 471, by permission of General Electric Co. and miSci, Museum of Innovation and Science.

as a site where the process of innovation, once conducted by individual researchers according to their own interests and agendas, could be organized, rationalized, and made more predictable and profitable.

The pursuit of x-ray technologies for plant breeding as an attempt to industrialize the process of biological innovation is most apparent in the efforts of these GE engineers. Their work might seem unusual in the history of plant breeding and remain unknown in the history of agriculture, but it sits comfortably with other narratives drawn from the history of technology—namely, the history of the industrial research laboratory, and the transition of innovative activities from the bench of the independent inventor to the well-outfitted company research laboratory, from unpredictable windfall to regularized production.[4] This case, I argue, helps us see the much larger ambitions driving research in x-ray-induced mutation, and the entanglement of plant breeders with the technological and industrial hopes of the mid-twentieth century.

It was not unprecedented, in 1930, for industrial producers to be interested in the development of agricultural technologies or even biological products.

There had been significant interest in the interwar years (and later, too) in the development of chemical products for agriculture, including fertilizers, herbicides, and pesticides. In the United States, this interest linked the large industrial chemical firms with agricultural scientists.[5] The chemurgy movement, which flourished especially in the 1930s, explicitly sought to forge an alliance between agriculture and industry through the innovation of industrial uses for farm products.[6] Undoubtedly, many in the 1930s understood that industry and agriculture could work to mutual benefit.

GE evidently agreed with this view. The x-ray mutation program at GE, which began in 1932, was described in one company report as a component of a larger initiative meant to bring applications of electricity to agricultural production. Most electric farm appliances on the market by the 1920s were laborsaving devices such as motorized milking machines and butter churns and an assortment of electric threshers, grinders, plows, and pumps. But as the executive research engineer Laurence Hawkins pointed out, the innovative energy that had been applied to these mechanical substitutions for the muscles of men, women, and beasts could be refocused on applications that enhanced rather than replaced living beings. Hawkins saw many possibilities, including artificial lighting to stimulate plants, electric heat for forcing the growth of plants, electricity for ionizing the air in greenhouses, and the application of x-rays to produce new varieties.[7]

Artificial lighting, electric heat, and air ionizers were areas of innovation that would benefit corporate welfare as much as they would human, at least within a company that had begun by producing electric lighting and power and subsequently pursued the development of electrical appliances for home, hospital, and industry. X-ray technologies, too, continued to be an important area of research and manufacturing for GE in the 1930s. Following the success of the Coolidge tube, it had consolidated its position in the American x-ray marketplace, in part by developing a midwestern company, Victor Electric, into an x-ray manufacturing firm (the Victor X-Ray Corporation) that would later become the General Electric X-Ray Corporation. By the 1930s, this company was handling about 50 percent of x-ray sales nationwide. If for no other reason, GE had an obvious interest in further development of x-ray technology and its applications.[8]

The company's research laboratory was the obvious place for these developments to be pursued. Such a goal entailed, in part, the continuous innovation and improvement of x-ray equipment. But administrators also considered it their responsibility to innovate uses of technology, in addition to instruments and machines. For example, in the immediate wake of the development of the Coolidge tube in 1913, research at the laboratory had emphasized the development and promotion of novel applications of x-rays in fields ranging from botany and zoology, to mineralogy and metallurgical research, to agriculture.[9] A few years later, Willis Whitney proposed that the company support a "radiographic institute" in Schenectady. He imagined that such a place would not only offer opportunities to test in practice the tubes manufactured by the company, but also provide information that would be helpful in advertising these for different purposes and possibly establish GE in life sciences research.[10] Although the radiographic institute never came together, Whitney did sponsor research into the hereditary effects of x-ray radiation through his encouragement of James Mavor's studies.

The agricultural work of the 1930s envisioned by Hawkins drew on these established patterns in the laboratory's innovation-centered research activities, both generating new technologies and devising novel uses for established ones. That did not mean, however, that the agricultural research was straightforward to carry out. Establishing an experimental program centered on the biological effects of x-rays meant first finding staff to run it and constructing new facilities. Because the laboratory staff comprised mostly engineers, physicists, and chemists (and not biologists), the program was run by employees diverted from other fields. One, Chester Moore, was a metallurgist whose previous contributions to the laboratory's output included innovations in electrical devices and x-ray equipment.[11] The second, Caryl Haskins, was a more recent hire. He was first assigned to work in plating and metallurgical processes before being redirected to the x-ray studies.[12] In order to facilitate the cultivation of many plant species, ranging from petunias to sequoias, a small greenhouse was constructed atop the main laboratory building and a field plot prepared nearby for growing the irradiated plants.[13]

The initial research carried out by Haskins and Moore centered on the assessment of x-ray effects in a range of crops, with the stated goal of advancing the use of radiation in inducing valuable mutations. They set out irradiating seeds, bulbs, and buds of many plants with very little knowledge of what the various tolerances of these plants would be, and therefore only a vague idea of how much radiation would be needed to create genetic changes. Limited by their meager backgrounds in biology, the two stuck to identifying gross morphological differences in the plants grown from irradiated material instead of carrying out genetic or cytological analyses. They hoped nonetheless to contribute to what research already existed in the area of x-ray-induced mutation, choosing plants that had not previously been given much attention, if any at all, and irradiating with high-energy or "hard" x-rays rather than the "soft" radiation that had been used by other researchers.[14] In the first few months, Haskins and Moore irradiated seeds of many species at varying intensities,

FIGURE 9. Gladiolus bulbs are prepared for x-ray treatment at the GE laboratory in 1932. GE Photographic Collection, miSci, Schenectady, New York, by permission of miSci, Museum of Innovation and Science.

searching for beneficial alterations. They treated seeds of typical crops such as cotton, peanut, radish, bean, turnip, and beet; trees including the tung tree, sequoia, and fruit-bearing trees such as lemon, lime, tangerine, grapefruit, and orange; and flowers such as aster, marigold, and still others. They also worked with bulbs of gladiolus, narcissus, and lily (fig. 9).[15]

Haskins characterized this effort as partly motivated by a search for specific improvements in economic plants, and claimed as inspiration the work of others in producing with x-rays such improvements as freestone cotton (which had been the outcome of Goodspeed's earlier work in California) and vine-type barley (produced through Stadler's experiments in Missouri).[16] But his and Moore's celebration of seemingly small changes should not be read as evidence of limited ambitions. The scope of change in agricultural production that they imagined could be produced through single-trait mutations was enormous, which in turn serves as a reminder that specific genetic mutations were thought of as highly valuable innovations. For example, Haskins asked one reporter to consider "the economic results of producing a cold-resistant orange." He envisioned winter-tolerant oranges, grapefruits, and lemons growing in the Northeast, even in Schenectady. Likewise, the irradiation of redwood and sequoia seeds might produce "a giant redwood adapted to the unfamiliar climate of New York, Illinois, or Colorado," thus transforming the national lumber industry.[17] These were hoped-for biological innovations that would reshape not only specific crop varieties but also landscapes of agricultural production across the United States. These grandest of the project's ambitions were no doubt highlighted for the benefit of a mass audience. Yet there is reason to believe that Haskins and Moore really did see them as potential outcomes of the research. American breeders had been working to improve the cold tolerance of citrus fruits at least since the 1910s, with some success, through straightforward hybridization and selection.[18] Why not assume that a genetic modification produced through x-ray irradiation would carry this work still further?

Haskins and Moore's research also tackled questions not overtly linked to plant breeding. From the outset, they expressed interest in the underlying mechanisms of x-ray-induced mutation, such as the physical processes by which radiation worked its effects in the cell. They soon began to investigate the amount of radiation required to "inactivate" mature pollen (in other words, to halt the growth of the pollen grain) as well as to similarly halt the growing capacity of mold spores. They studied the biology of flower color modification, the nature of somatic mutations in petunias, and the effects of radiation on the sex chromosomes of guppies, among other subjects.[19]

All these experiments, breeding-related or not, might well be understood along lines similar to those of earlier explorations of x-ray applications at the GE laboratory: they were efforts either to innovate new uses for an existing technology or to promote established uses. For example, if x-ray machines could be demonstrated to generate valuable mutations in important crop plants, then agricultural experiment stations, commercial seed producers, and ornamental flower bulb producers would be potential consumers of x-ray technologies, as physicians, dentists, and hospitals were already. The less practically oriented experiments, by comparison, did not represent efforts to innovate a novel use of x-rays—the irradiation of plants, molds, and flies was, as Haskins and Moore well knew, an established area of scientific experimentation. These studies instead can be thought of as attempts to produce novel findings within the general study of radiation effects, such as the effect of radiation on the inheritance of flower color. This research would contribute to the growing literature on x-ray effects and ideally call further attention to the uses of x-ray treatment in biological research. It would also contribute to the calibration of x-rays for use in various studies, enabling other researchers to choose an appropriate radiation dosage in their own investigations.[20]

This assessment of the purposes of the x-ray project within the larger corporate agenda of GE may be explanation enough for its existence. Yet to see this work as simply the further promotion of an established GE technology through the innovation, standardization, and publicizing of its applications would be to miss the more ambitious aims of this program. The project was also an attempt to develop x-ray devices into tools for industrial-style plant breeding through the continual, and hopefully predictable, radiation-initiated production of biological novelty. This aspect of the program bears the most significant implications for the history of biology and technology.

Although researchers had by 1932 been trying for five years to transform x-ray tubes into tools for agricultural breeding, the precise production of a specific genetic mutation through radiation treatment was still far beyond reach. As Haskins described in simple terms for a 1932 newspaper feature, although x-rays could "shatter" the pattern of genes, "the baffling feature of it all . . . is that we cannot know in advance what the new pattern will be."[21] This was true; none of the biologists then studying radiation-induced mutation could do anything but produce random effects. A key need was, therefore, to find a means to understand and predict x-ray effects, a challenge that the GE researchers readily claimed as theirs to tackle.

Although they remained vague about how exactly to resolve this challenge, they felt strongly that they were well positioned to do so. In Haskins's opinion, this research was particularly well suited to the professions represented among staff of the GE research laboratory. Geneticists, he believed, could use x-rays to induce variations, then apply the novel traits and their knowledge of the patterns of inheritance to create maps of chromosomes and genes. Such maps would be guides to future genetic research and potential aids to agricultural breeding. "But the realm of the geneticist ends with the power of the microscope," Haskins declared, suggesting that biologists were ill equipped to investigate the workings of the biological world beyond a certain range. When it came to understanding the workings of genes and producing genetic changes with radiation, "here we enter the domain of the research physicist and engineer." According to Haskins, the skills of the physicist were needed to reveal the mechanisms through which x-rays produced their effects on genes, and therefore the physical nature of the gene itself. The engineer would "initiat[e] tube design" to meet the physicist's specifications.[22]

In short, Haskins was proposing that it was not at all odd to have physicists and engineers carrying out genetic or agricultural research with x-rays, for only these professions commanded the right technical know-how. It would be their skills, moreover, and not those of geneticists, that would turn the heretofore-random production of mutations into a precision process. This would occur both through research into the nature of the gene—pursued especially though knowledge of the physics of radiation and biophysical responses to radiation—and through mechanical innovations that would make use of and extend that fundamental knowledge. Haskins expressed confidence that once physicists determined how radiation affected the cell, this knowledge combined with biologists' genetic maps (and an appropriately engineered x-ray tube) would lead to use of x-rays to reliably generate specific mutations. To this end, he and Moore included in their published research papers extensive details about their irradiation procedures.[23] They celebrated the fact that the petunia project had already produced results that indicated the desired predictability might be possible. The *New York Times* reported that the GE researchers had found that x-raying the branches of certain plants "produces a different type of flower, reproducible, the same dosage producing the same effect in a definite cycle."[24] "Someday," Haskins declared elsewhere, "we'll be able to eliminate the element of chance, will know just where to aim and what intensity and degree of bombardment to use to attain the exact result."[25]

Although Haskins focused on the intellectual background and technical skills of the engineer and physicist in describing the suitability of the greenhouse pursuits to his particular institution, he might equally well have pointed to the importance of the engineering culture and industrial outlook of the research laboratory as a whole. Other researchers who attempted work in x-ray breeding focused on the particular species with which they were most familiar: cotton breeders carried out trials on cotton, sugar producers looked at sugarcane improvement, and so on. Few looked to combine their research in these individual crops with a search for the fundamental principles that would make x-ray exposure a useful technique across all species. By comparison, the GE researchers were looking not only to create single biological innovations (e.g., cold-resistant oranges) but also to create or perfect a new process for generating such innovations ("some day, we'll . . . know just where to aim and what intensity and degree of bombardment to use"). Just as the laboratory itself was idealized as a mechanism for the continuous production of mechanical and other innovations that would eventually be patented by GE and form the basis of its economic success, the x-ray tube would be a tool for the reliable production of valuable biological innovations that would in turn generate profit. Speeding up evolution with x-rays would do more than improve individual crops: it would align the process of biological innovation more closely with the standards and aims of industrial mechanical innovation.

In spite of the great ambitions of the GE research in agriculture and biology, it never produced any significant changes in crop plants or timber trees, nor did it lead to a new technology for plant breeding. The lone marketable innovation to blossom from these activities was a new type of lily. In one round of x-ray treatments, Haskins and Moore had exposed to radiation an assortment of bulbs of a variety known as the Regal lily. From these bulbs, two

plants emerged that failed to shed their pollen, which happened to be a promising feature from a commercial perspective. Where the striking white petals
of these lilies typically become dusted with the flowers' yellow pollen, the
blooms of the aberrant non-pollen-shedding varieties reportedly remained
unstained from the moment they opened until they finally withered.[26] GE
moved quickly to patent the flower, soon dubbed the Roentgen lily, much as
it would any innovation that might have market potential. Moore (Haskins
had by then left the GE laboratory staff) applied for a patent in 1934, under the
recently established United States Plant Patent system.[27] This decision to patent further highlights the alignment of ideas about biological and other types
of innovation, in this case both at the laboratory and in American law, where
an argument had been made and won that plant breeders needed intellectual
property protection in their innovations just as any other inventor did.[28]

A patent on the new variety was granted in February 1936 (fig. 10).[29] In the
meantime, the biological research project at GE was called to a halt, leaving
the lily as its only notable product. Haskins later attributed the discontinua-
tion of the program to the continuing economic crisis in the United States, and
to the laboratory's difficulties in allocating resources among competing projects.[30] It is likely that diminishing interest in x-ray-induced mutations at other
more notable plant breeding institutions across the United States—resulting
from the overwhelming failure on the part of any researchers to produce improved types through the process—further contributed to the decision. Perhaps Haskins's and Moore's own increasing biological knowledge played a
role as well.

Although varied motivations lay behind the work of the x-ray breeders of the
1920s and 1930s, economic concerns were among the most influential. Many
who experimented with the process hoped to produce valuable biological innovations in particular species. This was true of Goodspeed, who expanded
his x-ray work to include other economically valuable crops; of the cotton
breeders Horlacher and Killough, who looked for "progressive" mutations that
would indicate hope for more-productive varieties; and of the GE researchers,
who focused their attention and their x-rays on organisms, such as oranges
and sequoias, that they felt would most dramatically alter the possibilities
of agricultural production in the northern United States. And it was true of
those unequivocally engaged in commercial development, such as Foster at
the Maui Agricultural Company or the delphinium breeder Reinelt.

Of the various species on which researchers attempted x-ray improvement,
horticultural novelties proved to be the only arena in which quick success
could be claimed: for example, Burpee's X-Ray Twins or GE's Roentgen lily.

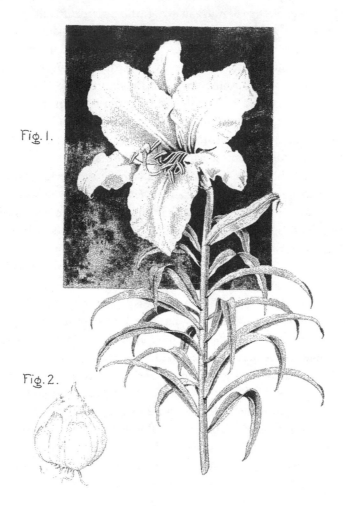

Fig. 1.

Fig. 2.

Inventor:
Chester N. Moore,
by *Harry E. Dunham*
His Attorney.

FIGURE 10. The US Patent Office issued this plant patent for a variety of Regal lily created through x-ray-induced mutation at the GE research laboratory. US Plant Patent 165, issued to Chester N. Moore and assigned to the General Electric Research Laboratory, 18 February 1936.

That it proved easier to produce a horticultural and chiefly ornamental "success" was related to the comparatively fewer constraints of garden or specialty-flower production and to the culture of novelty that pervaded it. Consider that the typical process of producing a variety for agricultural use in 1930 involved giving attention to a range of traits, many of which were difficult to measure, including size, vigor, yield, color, disease resistance, and suitability for certain climatic conditions, among other things. These traits would, moreover, need to be stable, guaranteed to appear in nearly all the individuals planted from a lot of seed so as to produce a reliable yield and a uniform crop. Hence breeders were justifiably concerned more with ensuring continuity in the appearance of traits, and not variability. Any new trait would have to be stabilized in order to be useful. By comparison, a variety of traits might also be desirable in a garden calendula and these would ideally be stable from generation to generation as well, but the economic stakes attached to achieving this stability would be much lower. For an American farmer, a corn crop lost because of a susceptibility to rust in the variety could mean devastation, especially during the agricultural depression of the 1920s and 1930s. In a gardener's backyard flower patch, far less was at stake.

In addition, the appeal of novelty—of double flowers, or new colors, or unusually shaped leaves—had long been a component of the production and consumption of ornamentals.[31] The usefulness of aesthetic variation in flowers in particular opened a whole new axis on which to evaluate the potential benefits of a mutation. Such changes could even be detrimental on balance, as, for example, in the case of variegated leaves (having less chlorophyll) or anthers that would not shed pollen (and therefore never sexually reproduce). In a garden or a greenhouse, however, even seemingly deleterious traits could be a competitive advantage if they appealed strongly enough to the sensibilities of the consumer. The commercial flower industry was just the right place to seek a biological-innovation machine, for there novelty as a feature of any given variety might trump even productivity or reliability. As a result, it seemed by the late 1930s that x-rays were suited to generate novelties for the gardener—and not much else. In the United States, only one crop variety bred from irradiated stock had been incorporated into large-scale cultivation by 1956: the Sanilac bean, a variety of field bean primarily grown in Michigan.[32] As a producer of valuable biological innovations, the x-ray machine was, by most accounts, a failed innovation.

One might well expect to learn that the dismal outcomes of nearly all the early efforts at x-ray breeding quickly squelched enthusiasm for further work in the field, and yet some researchers in the mid and late 1930s still insisted that it held promise. In 1934, for example, Goodspeed reaffirmed in his final report to the Committee on Effects of Radiation that induced mutations would be of practical use. The final committee report in 1934 arrived at much the same conclusion. The chairman Winterton Curtis expressed disappointment that the funding provided for radiation experimentation had not produced any spectacular breakthroughs in morphology, physiology, or genetics. He nonetheless maintained in his summary report that small and important advances had been achieved. Though a "thing of the future," Curtis concluded, "there is no doubt that eventually many such applications will be made in the heredity of domesticated plants and even in the heredity of man."[33] Even GE continued biological x-ray experimentation into the 1940s, with researchers such as the Cornell professor and plant breeder Bernard Nebel collaborating with GE scientists on the improvement of fruits and berries via x-ray irradiation.[34]

This continued attention to x-ray-induced mutation, and assertions of its potential for agricultural application, causes one to wonder: Why did the enthusiasm for x-ray breeding persist so long, when clear results were so few and far between, if they can be considered to have existed at all? There are many possible—and partial—answers to that question. All attest to the critical importance attached to the demonstration of x-ray-induced mutation, its perceived relevance to immediate human needs, and the alignment of the idea of "speeding up evolution" with broader commitments to efficient industrial production.

The most obvious answer lies in the fact that many people saw x-rays as the first proven tool for generating permanent inheritable changes in living things. Using radiation to produce mutations was not like taking advantage of the potential but unpredictable effect of the environment in causing a spontaneous mutation, nor was it planting and tracking many generations of a variety in search of the same. Although the exact mutations that would be produced through exposure to x-rays were unpredictable, that the organism and its future offspring could be made to change, could become plastic to human whims with just a zap of an electric ray, was taken as a dependable truth. At a time when the chemical substance and physical structure of the gene had yet to be determined, this effect could understandably be characterized as "magic," as it sometimes was in the popular press, for there was no truly satisfactory biological explanation. But it was also a real effect, one that produced visible alterations that persisted through many generations. As such, it had no comparable technology in 1930 save for exposure to radium, a far more expensive and less accessible tool.

It surely was also compelling that the imagined process of x-ray breeding would work by extending natural evolutionary mechanisms. However much reporters marveled at the idea and emphasized the extraordinary possibilities entailed on it, they did not describe the procedure of inducing mutations as a violation of the natural order. "Speeding up evolution" would not alter its course or lead to unnatural creations any more than artificial breeding already had. It would simply hurry things along. Mutations occurred all the time, in fields, forests, and farms, and these were well known to be the source of new varieties and theorized to be the starting point of speciation or at least the origin of genetic diversity within a population.[35] Connecting visibly altered traits to genic changes, and tracing the origins of these changes to a specific causative agent—radiation—spawned theorization on the role of natural radiation such as cosmic rays as an agent of evolutionary change via gene mutation.[36] X-ray irradiation was an artificial process, perhaps, but one with clear natural analogs.

Persistent interest in x-ray mutation also emerged from hopes that it was merely a first step toward the achievement of a far more selective improvement process. Time and again, researchers spoke of their aspiration to someday produce with x-rays a specific mutation instead of the inevitable random assortment of mutilated cells, chromosome aberrations, and recessive lethal genes. "When we know precisely how the x-ray photo operates within the cell to produce a mutation," mused GE's Haskins in one interview, "and when we know which genes control each type of mutation," then it will be possible to turn an existing plant variety into exactly the one desired.[37] Such aspirations freed researchers to see themselves as working on the improvement not only of a specific type of plant but also of the x-ray tube as a tool of genetic modification. This might have been a particularly influential idea for Haskins and Moore, working as they did in a context of industrial innovation where technological devices remained of surpassing interest. But it had not escaped even Stadler who had earlier made the same rationalization.[38]

Finally, the use of radiation to accelerate and potentially direct the appearance of mutations was of a piece with other trends in American agricultural production in the early decades of the twentieth century, which saw increasing emphasis on the adoption of industrial norms. Having a technology that would enhance control over the biological aspects of breeding efforts, and therefore ensure that these proceeded more efficiently and effectively, would surely contribute to greater agricultural productivity and help struggling agriculturists catch up with industrialists. It would streamline the process of plant innovation and improvement on which agriculturists depended. And this vision of radiation-aided plant innovation pushed still further, beyond the straightforward aspiration for industrialized agriculture. The aim was to make evolution itself more compatible with industrial ideals. The ability to precisely govern the appearance of mutations was envisioned as a way to turn the perceived slow and haphazard unfolding of evolutionary change (and not just the process of plant breeding) into a regularized, predictable, and efficient process. The appearance of new traits and therefore the production of wholly new types—the process of biological innovation—could itself be industrialized, just as other areas of innovation had been.

The persistence and visibility of attempts at x-ray breeding through the 1930s is therefore best understood as a convergence of many factors beyond just the potential for economic gain: namely, of this being the first, and for at least a decade the only, immediate means of altering inherited traits and therefore the best hope for a technology of genetic modification; of its perceived relationship to the processes of evolution and therefore the seeming ease and naturalness of its application; and of the promise it offered for more closely aligning the process of plant breeding, and therefore agriculture, with other areas of industrial innovation and production. These conditions would play a part in spurring similar research into the use of other mutagens as tools of breeding over the next three decades.

PART TWO

Tinkering with Chromosomes:
Colchicine in the Lab and Garden

In 1940, David Burpee, head of W. Atlee Burpee & Co., announced the release of what he called "the first new flower ever created by the use of a chemical."[1] Advertisements described this "Giant Tetra Marigold," a vigorous plant with thick stems and oversized blooms, as a "man-made miracle" and touted its unusual origins. "Giant Tetra is the result of doubling the chromosomes which govern heredity, in Guinea Gold Marigold," one such ad declared (fig. 11). Although some reports named Burpee himself as the "inventor" of the Giant Tetra, the plant had emerged from a series of experiments conducted at the New York State Agricultural Experiment Station in the late 1930s. There, two biologists had exposed seeds and shoots of various plants—among them, Guinea Gold marigolds—to a naturally occurring plant alkaloid called colchicine. When applied to plants in a solution, the alkaloid had the effect in a number of species of doubling the chromosomes, which in some cases generated visible changes, including larger, hardier blooms as in the marigolds.

The prospects of this chemical treatment thrilled Burpee, who characterized it as a way "to shock plants scientifically" so as to create new varieties. He celebrated it alongside x-rays, ultraviolet rays, deliberate mutilation, aging of seeds and pollen, chemical injections, and hot and cold treatments as ways for what he called "plant builders" to take control of the processes governing heredity. The alternative was to depend on nature to generate such shocks—"a hot day followed by a frosty night . . . or an aged seed, or aged pollen, or a mutilated plant"—and to hunt for the genetic mutations that might result. Burpee dismissed those as the "catch-as-catch-can methods" of yesterday and looked forward to a tomorrow in which he and his breeders "will work in a laboratory as scientifically equipped as that of a chemist or

Announcing the First Flower Created with a Chemical

Burpee's *Giant* Tetra MARIGOLD

GIANT Tetra is the result of doubling the chromosomes which govern heredity, in Guinea Gold Marigold, by using Colchicine, thus creating this new tetraploid Marigold. This man-made miracle has resulted in both greater size of blooms and more vigorous plants.

The large, carnation-like flowers are 3½ to 4 inches across, deep orange, of great substance, long-lasting in water. Thick, heavy stems, rugged foliage. Upright plants, 2 to 2½ ft. tall. Order now, and start indoors for earlier blooms.

Packet (75 Seeds) 25c
175 Seeds for 50c; 400 Seeds for $1.

Tetra Marigold, the newest American flower first shown on January 29th, at the Waldorf Astoria in New York City

FIGURE 11. W. Atlee Burpee & Co. advertises its new "Tetra Marigold," seeds of which were released for sale in 1940. As the ad declares, "Giant Tetra is the result of doubling the chromosomes which govern heredity." From *Flower Grower*, March 1940, back cover, by permission of W. Atlee Burpee & Co.

physicist, and our experiments will be systematically planned in advance." In other words, the days of hunting at random for a mutation or endlessly testing different hybrid combinations were over. The future lay in directed innovation.[2]

Historians writing about innovation have long recognized the importance of both trial-and-error and tinkering approaches. The classic example of trial-and-error (sometimes called hunt-and-try) methods is provided by Thomas Edison, who, in the absence of a theory to guide his choice of materials in a given invention, would try hundreds of substances, one after the other, until he found something that worked.[3] Tinkering, by comparison, is more difficult to characterize simply. It might best be described as open-ended experimentation with the various components of an object or system, whether to alter, repair, or improve it. In the history of technology, tinkering is most often used to describe the work of hobbyists and amateurs, especially those for whom the manipulation of mechanical or electronic devices was a leisure pursuit. Early automobile users who made modifications on their Model Ts, ham radio operators, and computer enthusiasts have all been classed as tinkerers.[4] Whether carried out by amateurs or professionals, tinkering is recognized as playing a crucial role in science and technology. Tinkering might manifest in the construction or repair of scientific instruments, the continual and often unplanned modification of an experimental setup, or the maintenance and improvement of mechanical equipment in high- and low-tech settings alike.[5]

Trial and error and tinkering have also been important in the history of plant innovation. Luther Burbank, to take one example, famously searched among many extant varieties of a particular fruit or flower in hopes of finding a trait he wanted to see in his new creations, and made many hundreds and thousands of crosses in his quest to produce the best possible hybrid. And although David Burpee might have dismissed Burbank's old-fashioned "catch-as-catch-can methods," the truth was that Burpee's "systematically planned" methods of tomorrow were little different. They included, for example, the application of x-rays, which were well known to produce random effects. As Burpee recognized, the best his breeders could do in most cases was to imitate the "shocks" that occurred in nature by various imperfect methods, all of which required endless fiddling to produce anything close to a desired outcome. In many respects, his "plant builders" were more like plant tinkerers. They played around with flowers and pollen, chemicals and radiations, not knowing exactly the route by which improvement would occur or what the end result could or should be. In the process, they tinkered with genes and chromosomes, too, hoping to stumble upon a fruitful assemblage of these that would result in a new variety.

Burpee even engaged his customers in these classic innovation activities. Beginning in 1954, Burpee asked gardeners to hunt among their flowerbeds for an all-white marigold (and offered a $10,000 prize to the first to discover one), something he desperately wanted to add to his catalog collection. Eight years later, still without a white marigold, he offered irradiated "Miracle Marigold" seeds for sale, promising that the radiation treatment "multiplies manyfold the chance of a new and different color, shape, or other change." In other words, the tinkered-with genes held greater promise for discovering the elusive all-white flower.[6] These were marketing ploys, and Burpee knew the market well. Many midcentury American gardeners were eager to try novel varieties and new chemical treatments. Much like other amateur experimenters and hobbyist technologists, they sought to make the latest offerings of science and technology their own. They trialed and tested, and tinkered with, their garden flowers.

Colchicine, famed progenitor of the Giant Tetra, promised to transform these methods, for professionals and amateurs alike. Described in the 1930s and 1940s as producing the same effect of chromosome doubling after every application, regular as clockwork, mastery of the technique was thought to make possible not only the deliberate creation of bigger flowers and similarly impressive fruits and vegetables, but also completely novel hybrid combinations. In short, the chemical promised to make possible a new engineering approach to breeding in which both trial-and-error methods and desultory tinkering, whether with techniques of mutation or genes themselves, would no longer be necessary. Instead, with colchicine, a new plant variety could be envisioned beforehand and then "built to order."

Various users of colchicine, however, Burpee breeders and home gardeners included, had different ideas about what exactly colchicine would be good for—and not everyone wanted or needed to see the old approaches disappear.

6

Artificial Tetraploidy

In 1925, the cytogeneticist John Belling proposed to readers of the *Journal of Heredity* a new method for the improvement of crops, fruits especially. The technique was simple: the application of ice to developing flower buds. Belling had observed that this cold treatment sometimes led to plants with unusual chromosome counts. Some appeared to be triploid, plants with three complete sets of chromosomes instead of the usual two, and others tetraploid, with four sets. He speculated that such changes might be useful if produced in the right plants. Triploids, for example, nearly always failed to set seed, and in plant varieties propagated by cloning such sterility could be valuable. Belling imagined the production of "seedless triploid apples, plums, peaches, mangoes, avocadoes, cherimoyas, etc.," fruits that might prove appealing to consumers as well as to commercial processors who would not have to trouble with removing seeds and pits. Tetraploids were desirable for another reason. These frequently were larger than their diploid counterparts, which in turn meant much larger fruits, flowers, and fibers. Belling suggested that creating tetraploid plants was a possible route to more productive cotton and flax, to sugarcane with enlarged stems, and to larger-growing timber trees, in addition to bigger fruits and ornamental varieties. "In fact," Belling concluded, "artificial tetraploidy seems to have before it a wide field in the improvement of the chief economic plants."[1]

Excited by these prospects, Belling and his fellow researchers at the Department of Genetics at Cold Spring Harbor soon undertook the construction of a "cold chamber" that would better enable them to study how temperature changes could be used to produce what they called "chromosomal types." The aim would be to bring the largely unpredictable process of chromosome change under far greater control, a feat that they assumed would have benefits for the production of economic plants as much as it would for genetics research.[2] The outcomes produced in the cold chamber would prove to be far less controlled and predictable than the Cold Spring Harbor researchers initially hoped—and yet they were pleased to have made some progress toward a technology for inducing chromosomal changes.

Much of the research on x-ray-induced mutation had been spurred on by the idea that the best way to intervene in heredity and evolution was to create change in genes, but as Belling's endorsement of "artificial tetraploidy" indicates, chromosomes were of equal and even greater interest to some plant biologists. Cytological analyses—in which slides were made from plant material such as pollen or root tips and stained so that the cellular contents (especially nuclear material) would be visible under a microscope—had suggested as early as 1907 that individual plants with unusual morphologies could sometimes be explained by their having more or fewer chromosomes than other individuals of the same species. Subsequent studies revealed consistent differences in chromosome number among closely related species of plants. Soon many plant biologists came to consider loss or accumulation of chromosomes, or the duplication of an entire set of chromosomes, to be a distinct mechanism through which speciation could occur. Still other studies revealed that the number of chromosomes among related varieties of economically important plants sometimes varied according to an arithmetic sequence (e.g., 8, 16, 24), suggesting that the addition of a complete set of chromosomes had occurred one or more times in their evolutionary past. This in turn indicated to researchers that induced "chromosomal mutation" and "artificial tetraploidy" had the potential to be useful techniques for agricultural and horticultural breeding. There was only one problem: no one had a good means of inducing such changes.

The unusual *Oenothera* plant holds no small responsibility for causing cytologists and geneticists to consider the possible role of chromosome number in evolution—and subsequently to devise ways of manipulating chromosome number in hopes of producing new types. Hugo de Vries's assumptions about the role of mutations in evolution, as illustrated through his observations of *Oenothera*, inspired others to investigate the unusual behavior of this "mutating" plant and in turn to link their findings with evolutionary explanations.[3]

One biologist whose research followed this trajectory was the botanist Reginald Ruggles Gates. He first encountered *Oenothera* in 1905, when he was twenty-three years old and just at the start of his graduate work in biology.[4] That summer, while conducting research at the Woods Hole Marine Biological Station, he had the opportunity to perform cytological studies on

Oenothera using seeds from some of de Vries's own plants. The initial goal of this effort was to investigate pollen formation, but his cytological work carried him in a new direction when microscopic slides prepared from pollen of the types *Oenothera lata*, *Oenothera lamarckiana*, and a hybrid of the two revealed differences in chromosome morphology and number. Gates speculated that the differences he observed in *O. lata* and possibly some of the other *Oenothera* mutations might result from an unusual distribution of chromosomes.[5] He subsequently launched a broader investigation of chromosome morphology and number in the named mutants of the species. The research revised but also defended de Vries's mutation theory. Gates maintained that de Vries's claim that the unusual *Oenothera* offspring were unique species was correct. The plants were not only easily distinguished morphologically but also, as his work demonstrated, had unique chromosomal configurations.[6] Gates found "irresistible" the conclusion, drawn from his analyses and the continuing hybridization experiments on *Oenothera*, that "all evolutionists in [the] future will have to reckon with" the "mutation phenomena" de Vries had identified.[7]

Gates and other biologists pursued the cytology of *Oenothera* vigorously in subsequent years. They demonstrated, among other things, that many of the morphological differences in various mutants named by de Vries resulted from chromosomal differences. For example, a common mutant of *Oenothera* named *O. gigas* for its large size had a chromosome count double that of its progenitor, with each cell carrying 28 chromosomes compared with the original 14.[8] This phenomenon of tetraploidy, in which a plant has twice the expected diploid number of chromosomes, had been observed by scientists in other species.[9] It had also been described by another researcher working on *Oenothera*. The biologist Anne Lutz had independently discovered in 1907 that somatic cells of *O. gigas* had about twice as many chromosomes as those of *O. lamarckiana*.[10] Lutz subsequently demonstrated that mutants of *Oenothera* known as *O. semigigas* were triploid in their cell makeup—instead of 14 chromosomes like the diploid *O. lamarckiana* or 28 like the tetraploid *O. gigas* plants, *O. semigigas* contained the intermediate 21 chromosomes.[11] Still other mutants of *Oenothera*, much like *O. lata*, were discovered to have one or two extra chromosomes. By 1920, about twenty of these had been catalogued, most of them by Gates and Lutz, providing evidence that some differences in plant morphology could be linked to differences seen among their chromosomes.[12]

The apparent significance of these chromosomal differences led Gates and others to speculate further on the role of spontaneous chromosomal changes in evolution. Gates believed that the mutants with an "unbalanced chromosome number" including *O. lata* and *O. semigigas* clearly demonstrated that mutations, such as those identified by de Vries, arose from alterations in the nuclear structure that were reproduced in cell division and therefore passed from one cell generation to the next. He also saw that these unbalanced types were unlikely to play a role in evolution. The unpaired chromosome was not passed on to every gamete, and as a result the plants did not often breed true. Tetraploidy, on the other hand, he judged to be "a condition of evolutionary significance." After all, the tetraploid plants demonstrated normal fertility and consistently passed on the tetraploid condition to their offspring.[13]

Gates was not alone in thinking that the doubling of chromosomes might play a role in evolution. By 1920, biologists had uncovered a range of plant genera in which species or varieties differed in chromosome number by a set factor, a condition suggesting that all the chromosomes had been duplicated one or more times. These included horticultural favorites such as roses and chrysanthemums, agricultural crops such as wheat and oats, and wild species such as maple trees.[14] Similar phenomena had also been discovered among animals, although the condition was much less common.[15] The observation of these many instances of apparent chromosome doubling (or tripling, or more) suggested the process might be a mechanism of speciation: it appeared to give rise to true-breeding types distinct from their immediate forebears. The Danish biologist Øjvind Winge further hypothesized in 1917 that the condition of tetraploidy could conceivably allow an otherwise sterile hybrid to become a new, fertile type. Although he had little evidence to back up his ideas, he speculated that hybrid sterility might arise from a failure of pairing between the chromosomes derived from different parent plants in the gametes of the hybrid. However, if the hybrid were to undergo spontaneous chromosome doubling, then this problem would be solved. There would be double copies of chromosomes from each parent plant, which would easily pair. The result would be a fertile hybrid, albeit a hybrid with a chromosome count that was twice the diploid number of the original pairing.[16]

While Gates pursued his work on *Oenothera*, another research team considered a similar set of questions using a different experimental organism. At Cold Spring Harbor, the botanist and geneticist Albert Blakeslee and his colleague Belling teased out chromosomal differences among many commonly occurring variations of *Datura stramonium*, the jimson weed plant. Their studies provided a case nearly identical to that discovered in *Oenothera* in which regularly occurring variants of jimson weed could be linked to chromosomal differences.

The odoriferous, hallucinogenic jimson weed plant was not a common experimental organism for a geneticist of the 1920s. Blakeslee had been working with jimson weed since the early 1910s, when, as caretaker of a small botanical garden at the Connecticut Agricultural College at Storrs, he received from the USDA a package of jimson weed seeds. The USDA official who had sent the seeds intended for them to be planted as an example of an "economic weed," but Blakeslee saw another use. He had been at Storrs since 1907 and served as both professor of botany and director of the summer school. He was the instructor for a number of subjects, including genetics. Flower color in jimson weed was inherited in clear Mendelian patterns, as was the form of the seed capsule (spiny versus smooth), making it an ideal organism for teaching Mendelian inheritance.[17]

Jimson weed first piqued Blakeslee's interest as an object of research when his assistant B. T. Avery took note of an unusual specimen that appeared in one of the station plots. The variation was soon nicknamed "Globe" after the shape of its seed capsule. Upon further study, it became evident that Globe did not pass on its distinctive seed capsule and other characters in the expected Mendelian ratios. These observed patterns of inheritance seemed to call for further analysis, but Blakeslee, busy with teaching and administration at the station, lacked time to follow up. The opportunity to make a closer investigation of Globe did not come until 1915, after Blakeslee had accepted a new position as a resident investigator in genetics at what was then still known as the Station for Experimental Evolution at Cold Spring Harbor. Since its opening in 1904, the station had supported research into the mechanisms of heredity and evolution. Station researchers deployed a range of approaches, from cytology to morphology to genetic analysis, and relied on a variety of organisms, from flies to maize to mice. Their work was funded by the Carnegie Institution of Washington and overseen by the station director Charles Davenport (who had also founded the Eugenics Record Office at Cold Spring Harbor in 1910, just a few years after the station opened). Blakeslee had spent time at the station during an academic leave from Storrs in 1912 and 1913. When a position in plant genetics opened at the station in 1915, Davenport offered him the post.[18]

Early in his tenure at Cold Spring Harbor, Blakeslee tried out a variety of experimental organisms for a new research program in genetics, eventually settling on jimson weed as the organism of chief interest. Together with an assistant at the laboratory, initially B. T. Avery who had come with him from Storrs but was soon killed in World War I and replaced in the lab by his younger brother Amos, Blakeslee began to grow large numbers of jimson weed plants at the station. The plant would remain central to Blakeslee's research for much of the remainder of his career. The chief reason for Blakeslee's consuming interest in jimson weed was that Globe turned out to be only one of many variant types that did not pass on their traits in the manner Blakeslee

expected. These variants, which Blakeslee called "mutants," occasionally produced offspring like themselves, but more often than not reverted to the common form. To Blakeslee, who claimed later to have been inspired by de Vries
and to have hoped to find for himself an example of a mutating species, jimson
weed must have seemed like a research goldmine.[19]

Blakeslee's jimson weed research began in the manner of a natural history survey, albeit one bolstered by his knowledge of genetics. He and B. T.
Avery compiled information on the thirteen "mutants" of the plant that had
appeared since Blakeslee had begun cultivating it. They noted the characteristic morphological features of each and puzzled over the fact that these were
not passed on in a Mendelian fashion but were rather inherited as an entire
suite of characters, which they called "complexes," by just a few offspring. The
exception to these unusual patterns of inheritance was a mutant they called
"N.S."—for "New Species"—which bred true and could not be crossed with
the apparently normal jimson weed plants from which it had originated. As
Blakeslee and Avery could not explain the origins of any of these curiosities,
the work seemed to arrive at a dead end.[20]

The jimson weed research, and especially investigation into the genetics
of the unusual variants, turned newly productive with the arrival of Belling
at the station in 1920. Belling had had a meandering career up to that point,
working variously as a science instructor, a botanist, and a horticulturist in
England, the colonial West Indies, and the United States. But he was, by all
accounts, an incredibly talented cytologist.[21] This skill he demonstrated in
his collaboration with Blakeslee during which the two quickly unraveled the
mystery of the unusual jimson weed plants. The key to their success was in
Belling's preparation of slides of the cell nuclei, the clarity of which enabled ac-
curate counts of the chromosomes of all the variants of the plant. In this task,
Belling's iron-acetocarmine technique for preparing slides was particularly
critical, as it revealed the chromosomes more clearly than perhaps any other
method of preparing cells for microscopic analysis at the time.[22]

Among Belling's first tasks at the station was to prepare slides showing
the chromosomes of jimson weed plants after the first division of the pollen
mother cells. These are cells with the normal diploid complement of chromosomes that undergo meiosis to eventually form the haploid gametic cells
of the pollen grains. The somatic cells of a normal jimson weed contain 12
pairs of chromosomes, for a total of 24. The slides Belling prepared from the
pollen mother cells of these plants after the first division of meiosis showed
that the gametes contained the expected 12 chromosomes. But the unusual
jimson weeds—the twelve identified variations plus the provisional New
Species—were different. In the twelve mutants, Belling's slides showed that

half of the time the gametes had 13 chromosomes instead of the expected 12. This meant that the mutant plants each had a somatic count of 12 pairs plus one extra, a total of 25 chromosomes instead of the expected 24. These were labeled "trisomics." The exception to this pattern was the New Species, which appeared to be a tetraploid, having 48 chromosomes in its somatic cells.[23]

The implications of these findings were clear to Blakeslee and Belling. The addition of an extra chromosome led to the unique suite of characters seen in each of the twelve variants. It could be assumed as well that differences among the twelve variants resulted from their each having an extra copy of a different chromosome.[24] As the jimson weed studies developed over the next several years, it became apparent that still other observed phenotypic variations—slight modifications of the mutants—could be explained by their having more complex arrangements of chromosomes involving multiple duplications. Some had, for example, two extra copies of the same chromosome or had duplicates of two different chromosomes. Still others were found to be missing chromosomes.[25] Among other things, this research demonstrated the potential rewards to cytogenetic analysis. One could identify specific chromosomes and observe their behavior during meiosis through the fixing and staining of plant materials on microscope slides during the appropriate point in the cell cycle. These observations could then be associated with data on inheritance gathered over many years in order to understand the origins of the unusual jimson weed plants.[26] Blakeslee was convinced that the many jimson weed variants would be useful tools for genetics research.[27]

These findings also led Blakeslee to speculate on the role of chromosomal changes in evolution, as Gates had after observing the effects of chromosomal change in *Oenothera*. Of particular interest to Blakeslee was the role of doubling the entire set of chromosomes—not surprising given that he had identified his tetraploid as New Species before he even knew why it differed from the other variants.[28] He knew that similar findings of tetraploidy had appeared in other plants grown in experimental cultures. *O. gigas* was the best example of these, but giant tetraploid variations had also been discovered in the flowering plant *Primula* by the botanist Reginald Gregory more than a decade earlier.[29] The tetraploid *Oenothera* and *Primula* were, like the tetraploid jimson weed, true breeding and incompatible with the parent plants from which they had been derived. Blakeslee suspected from this evidence that tetraploidy might be an important feature of speciation in plants. The evidence for the twelve trisomics being similarly significant was less compelling. These, after all, did not pass on their unique chromosomal arrangements with any regularity, and tended to interbreed with the parent generation. They could hardly be considered new species, at least in Blakeslee's assessment.[30]

It took only a small imaginative leap to move from speculating on the potential evolutionary significance of chromosome changes (that is, their likely role in speciation events) to considering their potential economic and agricultural uses. Blakeslee, perhaps encouraged by Belling but likely also drawing on his own experiences in agriculture, saw that these "chromosomal mutations" might constitute a new mode of plant breeding. It was clearly of "theoretical interest" to attempt the experimental production of chromosomal mutations. However, as Blakeslee mused in 1922, "it might also prove to be of considerable economic importance to be able to produce at will the full range of chromosomal mutants in any plants, especially in those which are propagated by vegetative means."[31] This, presumably, was because such chromosomal mutants might possess traits that would be advantageous in an agricultural or horticultural variety. If, as Blakeslee imagined, one could develop a means of controlling the otherwise spontaneous appearance of these mutants—such that they could be produced "at will"—breeders would have a powerful new tool in their hands. Most intriguing from an agricultural perspective was the production of triploids and tetraploids. Belling spelled this out in his *Journal of Heredity* article. The "true triploid" was chiefly distinguished by its being sterile, and therefore seedless. The "true tetraploid" was likely to be larger than its diploid parent. Both had immediate advantages for commercial growers: the production of seedless fruits and flowers in the case of triploids, and the production of enlarged varieties in the case of tetraploidy.[32] Even chromosome reduction could be useful. A haploid plant, with a somatic chromosome count equal to that normally seen only in gametes was first discovered among the jimson weed plants at the station in 1922, and similar plants appeared relatively frequently thereafter. These suggested a way to rapidly produce purebred lines, if a haploid could in turn be used to produce a diploid offspring.[33] Given the prominence at the time of pure-line breeding to produce hybrids, this would be no small achievement.

These notions—the usefulness of the unusual plants for pursuing genetics, the potential for an experimental study of evolution, and the hope for economic gain—led Blakeslee to a new goal. As early as 1921, he began to search for ways of artificially producing the chromosomal changes that had appeared spontaneously in his plant collections, the origins of which he could say nothing about. This project fit well with the ongoing research program at Cold Spring Harbor in which the director Davenport, best known for his role in the American eugenics movement, not only stressed the importance of studying heredity and variation as it occurred in nature but also supported projects intended to bring these under greater human control.[34] Working with C. Stuart Gager of the New York Botanical Garden, Blakeslee undertook an investigation of the effect of exposure to radium on chromosomes. The collaboration produced seemingly positive results quite early on in the form of an unusual jimson weed derived from a plant exposed to radium. Nonetheless, a full examination of the results of the radium experiments remained in process for six years. After more studies and cautious rewriting, the radium study was published in 1927, with the authors declaring that radium had produced chromosome mutation as well as gene mutation.[35] This work was quickly overshadowed by the more dramatic x-ray-mutation studies of Muller and others.

In the meantime, Blakeslee and his colleagues at Cold Spring Harbor pursued other initiatives, such as the cold chamber project. The work on cold temperatures and chromosome changes had been inspired by Belling's chance observations of a natural event. While studying the tropical legume *Mucuna* at the Florida Agricultural Experiment Station, he had seen—or, rather, he had produced visual evidence through his cytological analyses—that a sharp cold spell had led to the plants producing many diploid pollen grains. Those represented a doubling of the haploid number expected for a gametic cell. A similar event with a different plant species also occurred during his time at Cold Spring Harbor.[36] Cold spells, it seemed, could be the source of chromosome duplication in nature. It stood to reason that they might be easily reproduced in the laboratory by the application of ice, as Belling soon suggested.

This was not as easy as first imagined. Early experiments revealed that the temperature shifts needed to be monitored and manipulated more closely than the application of ice would allow. In 1922, Blakeslee and Belling hoped for "more accurately regulated cold-temperature rooms" so that the "production of mutants" could be controlled.[37] This ambition was achieved a few years later, when "a cold chamber with automatic temperature control" was installed at the station. According to a 1926 report, Blakeslee succeeded in producing varieties of what he called chromosomal types in four tries out of five during a first round of experiments using the cold chamber.[38] But experimentation with the chamber was subsequently abandoned (or, at the very least, went unreported), suggesting that in the end the technique was not as useful as they had hoped it would be.

Meanwhile, other plant biologists found themselves increasingly interested in the same technological capacity, wanting, like Blakeslee and Belling, to induce chromosomal changes on demand. It was evident from cytological studies that one explanation for why closely related species sometimes resisted hybridization was that they differed in chromosome number—a problem of immediate relevance to breeding.[39] This was true for a range of crops, from lesser crops such as berries and brambles to crucial economic products like wheat and tobacco. It was also recognized that in some crops, including wheat, an increase in chromosome number could be associated with certain advantageous characteristics.[40] As a result, the attention of many researchers was drawn to the study of chromosome number and especially to the possibilities of "artificial tetraploidy." Through the 1920s and 1930s, plant biologists actively sought tools that would let them alter the chromosome numbers of their experimental plants.

A significant breakthrough seemed to have occurred in 1932, when Lowell Randolph, a USDA cytologist stationed at the Department of Plant Breeding at Cornell, published details of a technique he had developed for inducing chromosome doubling in maize. Where Belling had advocated cold treatment, Randolph advised heat. Inspired by research on the effects of high temperatures on the chromosomes in somatic cells, he had begun a program of "high temperature experiments" hoping to produce tetraploids in maize. After finding that plunging the roots of maize plants in hot water for an hour produced some tetraploid cells, he devised a method for exposing developing zygotes of the plant to heat. The experimental setup consisted of encircling a growing ear of corn with an electric heating pad, a method by which he could expose the ear to an air temperature of 48 degrees Celsius for about an hour. The technique produced 29 tetraploid seedlings out of about 1,650 plants grown—not an overwhelming outcome, but not bad either when no other method had yet produced such consistent results.[41] Soon a colleague of Randolph's at Cornell, Ernest Dorsey, was hard at work trying to use the same technique in wheat and its relatives, hoping to produce fertile tetraploid hybrids of ordinarily sterile pairings. Dorsey was pleased with the results in which he successfully induced chromosome doubling in approximately one of every thirty tries, a frequency he described as "relatively high."[42]

Randolph summarized the broad potential outcomes of this research into induced chromosome doubling. If successful, it would "make possible . . . the experimental production of new polyploid strains, the duplication of the polyploid condition in existing species, the production of hybrids between species incompatible because of differences in chromosome number . . . and the production of fertile tetraploid hybrids between species whose normal diploid hybrids are sterile."[43] The summary reflected an assumed set of uses for induced chromosome doubling that was quickly becoming accepted truth.[44] It was a little surprising in light of the lack of evidence for some of these, especially the production of fertile hybrids out of sterile crosses along the lines envisioned by Øjvind Winge.[45] But given the increasing attention to polyploidy in plant genetics and evolutionary biology, and continued speculation as to its value in agriculture, scientists remained eager to discover methods and tools for inducing chromosome doubling to create tetraploids or other polyploid types.

Exposing plants to temperature changes continued to be the most frequently discussed method for achieving this well into the 1930s, but it was hardly dependable. Although the treatment could produce a doubling effect among treated plants, the ratio of affected plants to normal ones was often low, and the process unpredictable. Researchers had various other means, most of which were used simply because they seemed sometimes to work, not because the mechanisms behind their occasional successes were well understood. The Harvard geneticist Karl Sax, who had undertaken his own experiments using both hot and cold treatments of experimental plants, catalogued the various causes of polyploid cells (or diploid gametes, which could give rise to polyploid types, or irregularities in cell division) known by early 1937. In addition to exposure to temperature extremes, he listed species hybridization, tissue regeneration following grafting or other injury, application of narcotics and various chemicals, genetic factors, infection by a virus or attack by insects, osmotic changes in the cell, mechanical injury caused by puncturing or centrifugation, the use of x-rays, and the application of ultraviolet radiation.[46] Not all were viable options for generating chromosome doubling on demand. The observation of irregular meiotic division resulting from insect infestation, for example, could hardly be transformed into a consistent and straightforward technique. Even the more promising approaches were known to be effective only a small portion of the time. One trial of the method of generating chromosome alterations through centrifugation, proposed in 1935 by the Bulgarian geneticist Dontcho Kostoff, resulted in a handful of changed-chromosome types (reportedly including at least one doubled chromosome plant) from an initial centrifugation of "several thousands of tobacco seeds."[47]

It is no surprise, then, that guides to inducing chromosome doubling offered the whole gamut of known methods as potentially useful techniques. A 1936 manual published by the British Imperial Bureau of Plant Genetics on the "experimental production of haploids and polyploids" laid out the range available to the eager experimenter, who might be assumed to keep on trying until one produced the desired results: injection of chemicals such as chloral hydrate, centrifugation, producing tumors and callous tissue (which sometimes contained polyploid sectors), x-rays, chloroform, grafting, puncturing buds, and so on. The handbook acknowledged that none of these methods were perfect but suggested that they perhaps would be improved in due time: "The practical importance of haploids and polyploids in plant breeding is being quickly recognized, and it seems possible that their artificial production will be simply a matter of technique in the near future."[48]

It is clear from such reports that by the mid-1930s breeders increasingly wanted a foolproof technique for altering the chromosomal number of a particular plant, a technique that would neither require tinkering with capricious methods and tools nor limit them to only the most erratic tinkering with chromosomes. In the meantime, however, researchers made do with whatever they could make work.

Evolution to Order

Although geneticists and breeders by the mid-1930s commonly accepted that the ability to generate polyploid types by inducing chromosome doubling was of immense consequence, a reliable method remained elusive. This was the status quo until the summer of 1937, when several researchers, including Albert Blakeslee, learned of the effects of a plant alkaloid called colchicine on cell mitoses, and pioneered its use as a chemical tool for doubling the chromosomes of plants. Colchicine differed in many important respects from the other techniques known by that time to produce chromosomal effects. It appeared to double chromosomes consistently. It could be applied in a variety of ways and at different stages of plant growth. It seemed to be effective on a remarkable range of species. It was also easily obtained and relatively affordable. So the announcement that colchicine could be used to induce the duplication of chromosomes generated what one observer described as a "wave of optimism which seems to have swept through the ranks of plant breeders throughout the world."[1]

The discovery of colchicine-induced chromosome doubling was hardly a startling event. It came about when several individuals, primed to understand its value by their own research and by a long precedent of speculation about and experimentation on chromosome doubling, were introduced to a technique already in circulation among biological researchers that they subsequently developed and applied to new ends. Yet the very fact that colchicine seemed to fulfill a long sought-after desire—to be able to induce chromosome doubling on demand—meant that it was touted as a revolutionary finding by many working in the field.

Like x-ray-induced mutation ten years earlier, the chemical was seen as a means of speeding up evolution. But it was more than that, for it promised from the start to generate predictable changes in chromosome number.

For years, researchers who wanted to investigate the effects of polyploidy or other chromosomal changes in their experimental plants or breeding material had resorted to testing out one unreliable method after another. Now there was a go-to first option. The days of tinkering with various experimental approaches appeared to be over, and perhaps the days of old, slow, imprecise, or unpredictable breeding methods would come to a close as well: for if colchicine promised an end to fiddling around with different tools simply to induce chromosome doubling, it also promised an end to fiddling around with various combinations of genes and chromosomes in the hope of producing some useful agricultural or horticultural type. The promise of chromosome doubling lay in a more controlled mode of breeding based on the chemical manipulation of whole chromosomes, which some observers contrasted to the hit-or-miss process of genetic recombination that accompanied hybridization. In Blakeslee's mind, the colchicine technique created the possibility for what he called "genetics engineering," the directed synthesis of new species based on knowledge of their genetic makeup. Although the label "genetics engineering" did not catch on, the basic idea behind it was pervasive among those who first experimented with colchicine in plant breeding.

Having access to chromosomal variants, tetraploids in particular, was increasingly central to Blakeslee's research in the late 1920s and early 1930s. He used these plants (many of which appeared spontaneously within his collections) as tools with which to synthesize still other chromosomal types in a directed fashion, creating as he called it "new jimson weeds from old chromosomes." "Getting them [extra-chromosomal types] is a relatively simply process if we have a $4n$ [tetraploid] plant to start with," he wrote, for this tetraploid could be crossed with a normal diploid to produce a triploid. That, in turn, could be crossed again with a diploid to produce the entire range of trisomic plants.[2] Then, from these, myriad other chromosomal types could be obtained, sometimes through the breaking (and recombining) of chromosomes after treatment with x-rays and radium. Through elaborate hybridizations made with detailed knowledge of the characteristics of the chromosomes being combined, Blakeslee and his staff began to produce what he called "pure-breeding extra-chromosomal types." Unlike the original chromosomal variants he had discovered, these would breed true when propagated from seed—and thus arguably constituted laboratory-made species.[3] Certainly Blakeslee saw them as such (fig. 12). "Our types we have ventured to call artificial or synthesized 'new species,'" he declared in 1934. "Their greatest difference from species in nature appears to lie chiefly in the fact that we know their method of origin, having made them up to specifications, as it were, from a knowledge of the

SYNTHETIC "NEW SPECIES"
Figure 33

Morphologically these types are strikingly similar even though their chromosomes are different. Each has material equivalent to a 2·2 chromosome extra. Their chromosomal models are shown in Figures 31 and 32. They are more different from the form from which they arose than some of the described species in Datura. The quotation marks around the words "New Species" are in deference to taxonomists some of whom would confine their concept of species to forms which have arisen in nature and would exclude from the term these pure-breeding types which have been made up to specifications from a knowledge of the different parts of the chromosomes involved.

FIGURE 12. "Synthetic 'New Species'" of jimson weed created by Albert Blakeslee in his hybridizations of various chromosomal variants of the plant. He explained his use of quotation marks around "new species" as being in "deference to taxonomists" who would object to his labeling as true species his pure-breeding types—"which have been made up to specifications." From A. F. Blakeslee, "New Jimson Weeds from Old Chromosomes," *Journal of Heredity* 25, no. 3 (1934): 105, by permission of Oxford University Press.

different parts of the chromosomes involved."[4] To his own mind, he was not simply speeding along a natural process of evolution but rather directing the creation of types that nature might never have produced on its own.

As Blakeslee became engrossed in these efforts to de- and reconstruct jimson weed plants, members of his staff joked about the possible extinction of normal jimson weed due to "the indiscriminate use of x-rays and radium tubes on seeds and pollen." There was no great cause for alarm, as they maintained in a teasing laboratory memorandum, because Blakeslee's incomparable knowledge of jimson weed genes and chromosomes—evident in the ever-increasing complexity of the language he developed to describe the various chromosomal segments and the plants that possessed these—would enable him to remake the normal jimson weed through complex crossings:

> Dr. Blakeslee maintains that even if normals [$2n$ plants] were lacking they
> could be synthesized. "All we need," he said, "is a Judas [a specific chromo-
> somal variant] which is male sterile and heterozygous for Line 7 [?]. This then,
> provided it has 2 humps on the Echinus chromosome [the extra chromosome
> distinct to the 'Echinus' trisomic], we would female back cross to a triploid
> Spinach [a triploid of the 'Spinach' trisomic] in which the Polycarpic half of the
> Rolled chromosome [the extra chromosome distinct to the 'Rolled' trisomic]
> is attached to Microcarpic chromosome [the extra chromosome distinct to

the 'Microcarpic' trisomic]. Due to the fact that the B-ring [?] does not cause pollen abortion, ½ of the F2 offspring from this cross should be normals, providing that our theories are correct."[5]

This gentle ribbing of Blakeslee's proliferation of terms for specific chromosomal types among jimson weeds, as well as specific chromosomes and parts of chromosomes, attests to his increasingly arcane knowledge of his jimson weed plants. This knowledge was essential to his effective manipulation of the plants through otherwise typical hybridization techniques. Clearly, aggregating chromosomal material to create a novel jimson weed was a bit more challenging than Blakeslee's breezy published descriptions suggested. It nonetheless remained a promising avenue of research.

As Blakeslee and others in the laboratory moved in the direction of "synthesizing" new types according to deliberate (if complex) plans, having reliable techniques for processes such as chromosome duplication took on new importance. But such techniques remained elusive. Controlled and predictable "synthesis"—the goal of their efforts—still relied on the random appearance of chromosomal types within their cultures. Then, in 1937, colchicine landed in their laps. At the time, colchicine was already a well-established tool of biological and medical research. The toxic compound is derived from the autumn crocus, *Colchicum autumnale*. This crocus and products derived from it were long seen to have medicinal properties; their use in the treatment of gout, for example, reached back to ancient Greece.[6] French scientists first isolated the colchicine molecule in 1820, and European researchers in subsequent years sought to establish its toxicity on whole organisms—it was known to kill sheep and cattle that grazed on it in the field—as well as its pharmacological uses. After the turn of the century, scientists began to investigate its effects at the suborganismal level as well. This research heated up in the early 1930s, when a pathology research group at the University of Brussels began extensive experimentation with colchicine. One member of the group demonstrated that the chemical induced the proliferation of mitoses in 1934. Then the head of the research group noted the ability of colchicine to produce what would become known as "metaphasic arrest"—a slowdown or stop in cell division at the point where the chromosomes have aligned within the nucleus of the cell immediately prior to its division (i.e., the point in the cell cycle known as metaphase). Studies produced by members of this lab and by a few other European researchers suggested that colchicine would be a valuable tool to use in the study of nuclear and cell division, for it was effective in both plant and animal cells and apparently enabled the researcher to exert some control over these basic cellular processes.[7] Soon word of the chemical's cellular effects reached both the geneticists working at Cold Spring Harbor and the cytologist Bernard Nebel of the New York State Agricultural Experiment Station in Geneva, New York.[8] It was at that moment that attention turned to another striking effect of colchicine: its apparent production of cells with twice the expected number of chromosomes.

At Cold Spring Harbor, an initial demonstration of the mitotic effects of colchicine by one of the resident physiologists inspired Orie Eigsti, a recent graduate who worked as a cytologist at the lab, to obtain colchicine with which to pursue his own investigations.[9] First Eigsti applied a solution of colchicine to root tips of onion. When the roots showed the development of polyploid cells, he turned to testing the solution on seedlings of radishes and corn.[10] These experiments, too, suggested that colchicine would be effective as a method for inducing chromosome doubling in plants, an effect apparently not previously sought by researchers who had used the chemical.

Prompted by Eigsti's independent study, Blakeslee and Amos Avery also began to work with colchicine, reportedly as part of a larger investigation of chemical means to induce chromosome doubling begun in June 1937. Of the substances they tested, which included chloral hydrate and "several narcotics," only colchicine showed the desired effect. As Blakeslee and Avery described, they had first soaked seeds of jimson weed in a solution of colchicine and tap water. A number of these seeds produced plants whose leaf and flower morphology suggested that they were tetraploid rather than diploid as the parent seed had been. These promising results led to a rapid expansion of the investigation into colchicine.[11]

The report of their experiments suggested that finding out how to make colchicine produce the desired effect required Blakeslee and Avery to make many trials, using different strengths of solution and different experimental organisms. The two played around with various methods of application (fig. 13). They immersed twigs and branches in solution, and applied colchicine to plants using agar solutions or via a smear of lanolin (also known as wool wax or wool fat). They tried targeted delivery of the chemical via capillary action along a string, placed single drops directly on the bud of a plant, and misted developing plants with a fine spray from an atomizer. Hoping to demonstrate that the effect was universal, they incorporated other species into the study. When they published their first full report in late 1937, Blakeslee and Avery described having successfully induced the doubling of chromosomes in at least twenty species using the various methods they had devised.[12]

Blakeslee announced the results of these experiments at a meeting of the Genetics Society of America in the autumn of 1937. He evidently was well aware of the fact that his priority—questionable to begin with given the Eu-

METHODS OF APPLYING COLCHICINE

FIGURE 13. Albert Blakeslee and Amos Avery experimented with a number of different methods for applying colchicine: dipping whole shoots in solution, spraying plants with an atomizer, applying the solution to buds with an eyedropper, using capillary action along a string, and covering growing points of plants in an agar solution of the chemical. From Albert F. Blakeslee and Amos G. Avery, "Methods of Inducing Doubling of Chromosomes in Plants by Treatment with Colchicine," *Journal of Heredity* 28, no. 12 (1937): 398, by permission of Oxford University Press.

ropean research and also that of Eigsti in his own lab—was at risk.[13] His haste to secure priority in publication upset Eigsti, who saw the senior scientist scooping his work and taking all the credit for recognizing the importance of the finding. Eigsti left the station soon after, though he continued his own studies with colchicine.[14]

During this same period of time, a husband-and-wife team at the New York State Agricultural Experiment Station undertook an independent study of colchicine and its effects on plant cells. Bernard Nebel, South African by birth and educated in Germany, had arrived at the experiment station in New York around 1930 to take up a position as research associate in pomology, the science of fruit growing. Nebel was an expert in cytology, though he also conducted research on plant morphology and physiology while at the station.[15] Mable Ruttle, who had a PhD in biology from Cornell, worked as a research assistant in botany at the station, sometimes in collaboration with her husband. Nebel and Ruttle began their experiments with colchicine in April 1937, prompted by reports of its effects they had learned from the geneticist and plant breeder Donald Jones. They aimed at first to describe the effects of colchicine on mitoses, but they seem to have redirected their research when they discovered the production of polyploid cells to be a common result.[16]

In their subsequent investigation of the potential effects and uses of colchicine, Nebel and Ruttle immersed parts of plants and in some cases whole seedlings in a colchicine solution, and "painted" portions of plants with colchicine that had been mixed in lanolin. They and their collaborators used a range of plants and plant parts to study the cytological mechanisms and physiological outcomes of the colchicine treatment. Unsure of how widely the effects applied, they tested it on the flowering plant *Tradescantia*, often used in biological research; on crop plants like maize and tomatoes; and on garden flowers such as marigolds, snapdragons, poppies, pinks, and lilies. They considered its effects on animals as well—grasshoppers, starfish, and sea urchins.[17]

Blakeslee and Avery's announcement in September 1937 undoubtedly pressed Nebel and Ruttle to declare the findings of their own research efforts. Their initial summary, published after a few months of experimentation, drew conclusions similar to those of Blakeslee and Avery. As they described, colchicine could be used to create polyploid cells in both plants and animals; in plants it could be used to produce entirely polyploid organisms in which duplication affected the gametic cells, though they noted that not every species was equally susceptible.[18]

The first round of publications on colchicine-induced chromosome doubling—by Blakeslee and Avery, Eigsti, and Nebel and Ruttle—called attention to the potential of colchicine to advance scientific research via its use as a tool in cytology and genetics. Eigsti noted that colchicine did not produce effects at random but rather demonstrated a "specificity of a high degree for inhibition of certain phases of cell division" while leaving other processes unaffected.[19] The implication was that it could be used to more easily explore aspects of cell functioning. Nebel and Ruttle emphasized the production of chromosome doubling in animals and plants, and the extent to which this could be accomplished through application of the chemical. This ability, they soon claimed, "realize[d] a long-cherished dream of cytologists" to easily make a range of polyploid types for further study.[20] Blakeslee and Avery, by comparison, made a case for colchicine's usefulness in what they called theoretical genetics. They suggested in the first instance that varieties with a wide range of chromosome combinations could be bred in almost any species, to produce types similar to the chromosomal types they had been able to breed in jimson weed.[21] These "unbalanced types," which the Cold Spring Harbor group already used to analyze the genetic makeup of jimson weed, could be produced to study genes and patterns of heredity in other species.

In their discussions of colchicine as a research tool, all the authors emphasized the apparent control conferred by the chemical treatment, that is, the way in which it promised to make it possible for geneticists and cytologists to produce specific changes in plant chromosomes. And as they well knew, this very capacity would also make colchicine an incomparable tool for plant breeders. Blakeslee and Avery pointed out the range of options in plant breeding that colchicine-induced polyploidy might make possible. For example, extra-chromosomal types could be used to build up purebred stocks, or they might be valuable new types in and of themselves. In addition, and as a particularly dramatic comparison of haploid, diploid, triploid, and tetraploid jimson weed flowers demonstrated, a plant's flower (and fruit) size sometimes increased along with its chromosome number—a long recognized feature of polyploid plants that emphasized their potential for increased productivity (fig. 14).[22] Perhaps most important of all, Blakeslee and Avery described how practical breeders might use colchicine to control more precisely the development of new plant varieties by producing fertility in widely crossed hybrids that were ordinarily sterile. The implications of such options were clear: "with increasing knowledge of the constitution of chromosomes and of methods whereby their structure and behavior may be altered, there arises an opportunity for the genetics engineer who will apply knowledge of chromosomes to building up to specification forms of plants adapted to the surroundings in which they are to grow and suited to specific economic needs."[23] For Blakeslee, having a technique like colchicine-induced chromosome doubling meant a

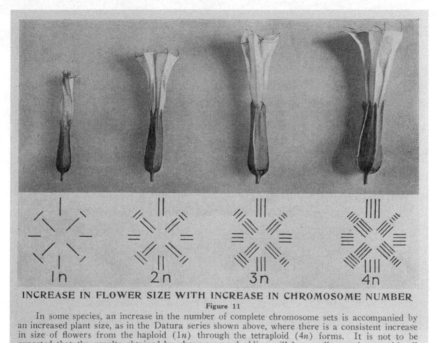

INCREASE IN FLOWER SIZE WITH INCREASE IN CHROMOSOME NUMBER
Figure 11
In some species, an increase in the number of complete chromosome sets is accompanied by an increased plant size, as in the Datura series shown above, where there is a consistent increase in size of flowers from the haploid (1n) through the tetraploid (4n) forms. It is not to be expected that the results obtained by chromosome doubling will be equally consistent with all species or always lead to varieties of economic value.

FIGURE 14. An increased number of chromosomes corresponds to an increase in flower size, at least in the case of jimson weed. From Albert F. Blakeslee and Amos G. Avery, "Methods of Inducing Doubling of Chromosomes in Plants by Treatment with Colchicine," *Journal of Heredity* 28, no. 12 (1937): 408, by permission of Oxford University Press.

new mode of plant breeding, one no longer dependent on the uncertain procedures of searching for sports or testing various hybrids, but guided instead by precise knowledge and precision tools.

Nebel and Ruttle agreed, asserting that in some cases colchicine methods constituted a wholly new approach to plant breeding. "To induce chromosomal changes in somatic tissues of cultivated plants as obtained with colchicine may be called plant breeding with non-Mendelian methods," they declared.[24] Unlike Mendelian breeding, which focused on individual characters and how the most advantageous combination might be produced in a single line of a given species, colchicine breeding involved the manipulation of whole chromosomes and potentially the combination of varieties that could not ordinarily be hybridized. Though colchicine had so far worked only with herbaceous plants, Nebel and Ruttle maintained that "there is no reason to doubt that the principles . . . are [equally] applicable to the fruits or that they [will have] equally far-reaching commercial consequences."[25]

The striking effects of colchicine, along with the idea that these might entail "far-reaching commercial consequences," proved an appealing news story. In the 1930s, demand for science reporting was on the rise, and news outlets were eager for accounts of startling discoveries or new technologies that would improve everyday life.[26] In such a market, reports of colchicine research, and speculations about the wonders it would soon generate, proved an easy sell. These reports tended to present the chemical as a straightforward means of controlling evolution and making new plants to conform to some imagined ideal—in short, as a tool for engineering life. "New control over fundamental processes of life in plants and animals—which may even give man power to direct the course of his own future evolution," declared one Associated Press story.[27] The outcome of this control would of course be "new, better, bigger and hardier flowers, fruits, vegetables and perhaps trees"—at least that was the tally offered in one *New York Times* report about the "new elixir" for improving plants.[28] A couple of months later, another AP report declared the likelihood of a "revolution on the farm," promising that a few colchicine-treated onions, radishes, and clover represented "the next great revolution in agriculture." Scientists had "discovered the double chromosome miracle" that would make ordinary crops larger and hardier than ever before.[29]

Some news reporters were willing to carry assessments of colchicine's value to obvious extremes. A case in point is an article that appeared in various Hearst newspapers in November 1937. The fantastical story—illustrated by, among other things, one woman pushing an oversized baby in a stroller the size of a car and a second under attack by a six-foot caterpillar—claimed that the "elixir of growth" colchicine could lead to giant insects and animals in addition to spectacular new plant varieties (fig. 15). "With colchicine farmers might grow wheat as tall as pine trees, but if they did they would have to raise their sons on the same stuff, or there wouldn't be anybody big enough and strong enough to harvest the crop," the Hearst reporter pointed out; he also speculated on the problems that scientists might cause if they accidentally spilled the chemical on a carnivorous plant, thereby creating a man-eating one.[30] Blakeslee first griped, and later joked, about the inquiries prompted by news stories like this one.[31] In later interviews, he stressed that colchicine would not, as the article had led some to believe, grow hair on bald heads, make short men taller, or create a superhuman race.[32]

Of course, Blakeslee was in large part responsible for these earlier exaggerated news reports, to which his enthusiastic self-promotion had directly contributed. He complained to fellow scientists about how colchicine was covered in the news, professing to be unhappy with what he considered journalistic hyperbole. He claimed to "fear that the effect . . . is to encourage people in

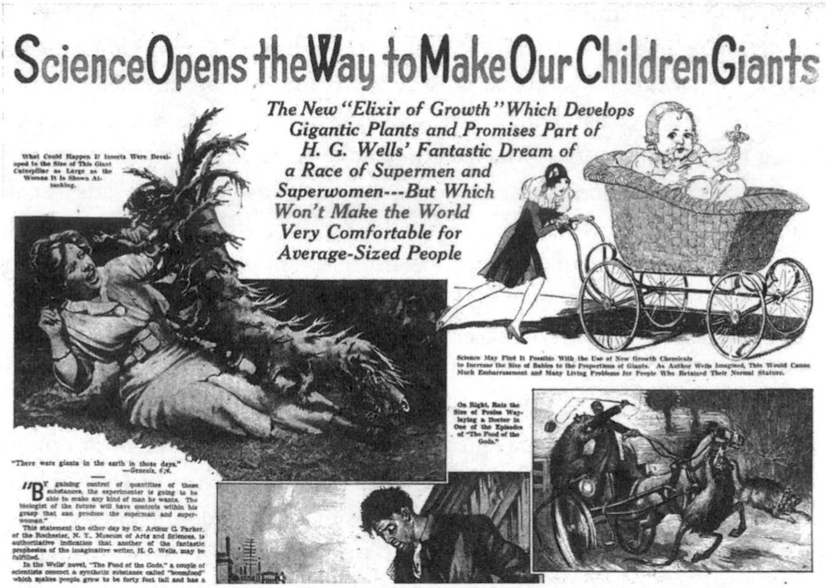

Science Opens the Way to Make Our Children Giants

The New "Elixir of Growth" Which Develops Gigantic Plants and Promises Part of H. G. Wells' Fantastic Dream of a Race of Supermen and Superwomen---But Which Won't Make the World Very Comfortable for Average-Sized People

What Could Happen If Insects Were Developed to the Size of This Giant Caterpillar as Large as the Woman It Is Shown Attacking.

Science May Find It Possible With the Use of New Growth Chemicals to Increase the Size of Babies to the Proportions of Giants. As Author Wells Imagined, This Would Cause Much Embarrassment and Many Living Problems for People Who Retained Their Normal Stature.

On Right, Rats the Size of Ponies Waylaying a Doctor in One of the Episodes of "The Food of the Gods."

"There were giants in the earth in those days."
—Genesis, 6:4.

"BY gaining control of quantities of these substances, the experimenter is going to be able to make any kind of man he wants. The biologist of the future will have controls within his grasp that can produce the superman and superwoman."

This statement the other day by Dr. Arthur C. Parker, of the Rochester, N. Y., Museum of Arts and Sciences, is authoritative indication that another of the fantastic prophesies of the imaginative writer, H. G. Wells, may be fulfilled.

In the Wells' novel, "The Food of the Gods," a couple of scientists concoct a synthetic substance called "boomfood" which makes people grow to be forty feet tall and has a

FIGURE 15. A sensational 1937 report on colchicine, "the elixir of growth," suggests that its successful use on plants portends a future of huge animals and insects—and maybe even human babies increased "to the Proportions of Giants." The article appeared in Hearst newspapers across the country as a feature in their Sunday magazine supplement. From *American Weekly*, 21 November 1937, 7.

the idea that by some wizardry we can produce big pumpkins out of a hat by impregnating the surrounding air with a magic mist."[33] And yet he was glad to give interviews to reporters and to participate in public outreach, where his own claims were not so different. Invited to participate in a March 1938 episode of *Adventures in Science*, a regular radio program produced by the science news agency Science Service, Blakeslee joined the host in a discussion of the topic "evolution to order." The host led off by noting, "I don't know whether science will ever produce giant dogs and cats or human giants, to order." But he did think it likely that there would soon be "bigger roses for the garden, and bigger tomatoes for the dinner table" resulting from research with colchicine. Blakeslee agreed with this prediction, assuring listeners, "Yes; we can probably make them up to order if you want them."[34] As he had in the *Journal of Heredity*, Blakeslee staked a claim for the engineering future of genetics. There would no longer be a need to test out many different hybrid combinations or tinker with unpredictable methods of altering genes and chromosomes in hopes of producing the desired traits or types. The future would see organisms built to specification—or so he imagined.

Other researchers' high hopes for colchicine were similarly captured in the press. A 1938 AP report, which began with the claim that colchicine had

"opened up possibilities in the plant world as far reaching as the introduction of the steam engine in transportation and industry," also presented Nebel
making bold predictions. In his (reported) words, through the use of colchicine, "vegetables, cereals and fodder plants, with due effort, may be changed
during the next 20 years as much as they have been transformed in the past
200."[35] A subsequent newspaper report quoted him as predicting "large-scale
commercial applications of this new method of regulating plant production."
The author offered proof that this was not too far-fetched an idea with descriptions of the "new types of plants, never before seen on earth" that Nebel
and his colleagues had produced at Geneva, such as "spearmint with a lemon
flavor" and "bigger and more rugged marigolds and snapdragons."[36]

Attuned to these potential benefits, and perhaps wanting to make good
on the claims they and others had advanced in their articles and interviews,
all the colchicine researchers sought applications for the new technique. The
official line heard by 1939 at Cold Spring Harbor was that despite the evidence
for the wide application of colchicine, the "prime interest" among those at the
station was "in its use as a new tool of research in major problems of genetics
and evolution."[37] But Blakeslee was determined to produce some striking proof
of colchicine's power. In correspondence with Arlow Stout of the New York
Botanical Garden, he proposed developing tetraploid poplars whose enlarged
size would make them better pulp producers and inquired into the usefulness of triploid grapes as seedless varieties.[38] He also attempted to establish a
cooperative colchicine breeding program with the USDA. "I do not need to
tell you about the practical as well as the theoretical importance as a tool of
research of a technique which would enable one to double chromosomes at
will," he wrote to Frederick Richey, the chief of the USDA's Bureau of Plant
Industry, in pursuit of his opinions on how best to develop the "practical"
side.[39] Richey, however, was not keen on cooperation.[40] A few months later,
Blakeslee tried a different route, writing to the Office of Drug and Poisonous
Plant Investigations, to see if the USDA researchers there might be interested
in a cooperative effort. Blakeslee speculated that doubling the chromosomes
of plants valued as sources of particular chemical compounds might double their production of those compounds. If tetraploidy could be induced in
mints, for example, then the content of menthol might increase along with
the size of the plant, making each mint plant more productive and therefore
more profitable.[41] This, too, proved a dead end.[42] Blakeslee had greater success
in inviting other collaborators. In November 1937, he began investigating the
potential application of colchicine in animal breeding. With the assistance of
researchers from the Cornell College of Agriculture and the superintendent of
the Cold Spring Harbor State Fish Hatchery, Blakeslee attempted to produce

"bigger and better" tetraploid fish through the use of colchicine on fertilized eggs.[43] This project and a similar experiment attempted on frogs were ultimately unsuccessful.[44]

In these efforts, especially those with plants, Blakeslee's goal was to transfer to practical breeding some of the ideas and techniques for altering chromosome configurations that he had developed in his work on "synthesizing" jimson weeds. "With the ability to double chromosomes it should be possible to make available to practical purposes the chromosomal findings in Datura to an extent that was not possible before this technique was discovered," he wrote to Richey at the USDA. He continued, "I feel very confident that in time various agricultural species will be studied from somewhat the same chromosomal standpoint . . . and that the extra chromosomal material will be consciously used in building up desirable economic varieties."[45] Blakeslee did not want to simply produce more positive results in his experimental jimson weed plants. He wanted to generate evidence of the usefulness of colchicine in producing agricultural crops and other economic products.[46]

Meanwhile, in New York, Nebel and Ruttle also searched for ways to put colchicine to work in breeding. They had been more cautious than Blakeslee in their initial analysis, emphasizing that the approach would "require more care on the part of the practical breeder and more expense than the Mendelian method" if it were to be successful.[47] But the agricultural experiment station offered them ample space and resources to pursue this non-Mendelian method of plant breeding. They continued to work with some of the flowers that had responded well in their initial experiments, especially marigolds and snapdragons. They treated the shoots of six different snapdragon varieties, the seed of which produced, in their estimation, "six new and handsome tetraploid varieties." These they judged to be superior commercial varieties: the flowers were larger and ruffled, with colors "stronger and deeper" than their diploid counterparts, while the plants overall were taller and sturdier.[48] Marigolds responded similarly.[49] Nebel and Ruttle also used marigolds to illustrate the usefulness of induced polyploidy in making crosses between species previously resistant to hybridization. They doubled the chromosomes of a marigold hybrid that ordinarily produced a sterile triploid, turning it into a hexaploid. The plant was more fertile than the triploid progenitor (also "larger" and "sturdier").[50] These and other floral victories, in Nebel's estimation, showed "that where its application has been mastered the drug [colchicine] fulfilled expectations."[51]

In addition to the work with flowers, Nebel and Ruttle were eager to develop colchicine as a tool for breeding crops, fruits in particular. Fruit improvement had long been a specialty of researchers at the Geneva experiment station, and Nebel, with his experience in fruit cytology, envisioned from the outset many benefits of colchicine in this area.[52] There was reason to believe that common fruits such as apples, pears, cherries, plums, and grapes, as well as many types of nuts and berries, could be improved via chromosome manipulation. In 1938, Nebel and Ruttle maintained that their initial successes with flowers portended longer-term achievements in many of these.[53] The work would be harder, for they quickly discovered that many fruit species were not as receptive to colchicine treatment as their flowers had been. This led them to begin tinkering with their methods again, in hopes of generating better results. They found, for example, that "the fruit buds must be pre-conditioned for effective treatment," and thus they built a cabinet "in which light, temperature, and humidity can be regulated and automatically maintained over a wide range."[54]

Nebel and Ruttle imagined that there would be a significant payoff for such troubles. "The peach, apricot, and 16-chromosome plums could, with all likelihood, if their chromosomes are doubled, be eventually intercrossed, and from this amalgamation new kinds of fruits and even new industries might arise," they speculated. They advocated the development of new triploid apples and pears, and of tetraploids of cherries, some grapes, and brambles such as raspberries and blackberries, where "at some future time breeding with diploids will have reached its natural limits."[55] Like Blakeslee, Nebel and Ruttle were also interested in the effect of chromosome doubling on the chemical content of various plants. Accordingly, they investigated the effects of chromosome doubling on herbs known for their oils, including basil, mint, and spearmint.[56]

In all their applied projects, Nebel and Ruttle were eager to prove that their cytological discovery could meet the expectations long anticipated to arise from the doubling of chromosomes. They sought the immediate advantages of polyploid types—namely, bigger or improved flowers and fruits, and possibly more oils—and also to produce fertile crosses from pairings that typically produced sterile offspring, and to cross species previously uncrossable because of their differing chromosome counts.[57] The advantages of colchicine over other methods of plant breeding seemed obvious. Like radiation, a technique that Nebel also used in efforts to induce variation, it offered the opportunity to circumvent the seeming limitations on the material offered by nature when breeding new varieties. However, unlike radiation, there also appeared to be a possibility that through colchicine breeders would, as Ruttle and Nebel wrote, "find ways of inducing changes in plants which can be more definitely predicted."[58] With colchicine, the outcome seemed both constrained to chromosome duplication and more reliable in its occurrence, meaning that plants might be produced according to a preconceived plan.

In the meantime, as their continued efforts to make the technique work in the most enticing cases makes clear, tinkering remained essential—even to work with this supposedly predictable new chemical method.

Better Evolution through Chemistry

The research carried out by Eigsti, Blakeslee and Avery, and Nebel and Ruttle in 1937—and especially the sweeping statements they and others made about the future possibilities it enabled—led to an avalanche of research on induced chromosome doubling and especially the application of colchicine to important agricultural and horticultural crops. Researchers had long understood induced chromosome doubling as a potentially valuable means of manipulating plants. As a result, there were numerous projects for geneticists and breeders to undertake using the new technique, many of which had been envisioned, and in some cases attempted, in the preceding decades. The ensuing "fad" for colchicine experimentation lasted well into the 1940s.[1] Some scientists found the faddishness irritating, especially the strong hold it appeared to take among American researchers.[2] Still, interest in colchicine and polyploidy persisted. By 1947, ten years after the initial journal articles, there were enough publications related to the colchicine technique that a bibliography produced for the botanical journal *Lloydia* ran to fifty pages.[3]

In some ways, this response was similar to the flurry of research into the genetic effects of radiation that followed the demonstration of x-ray-induced mutation. Geneticists and cytologists aspired to elucidate the mechanisms through which the alkaloid influenced cells. Accompanied by plant breeders, these researchers sought to determine the range of species on which it was effective and the ways in which a tool for doubling chromosomes could be made useful. But where the use of x-rays to produce mutations had sparked myriad claims about the ability of scientists to control evolution and engineer new organisms to order, the news about colchicine easily redoubled these. In replicating an important evolutionary process, the chemical was thought to accelerate the production of variations that would be potentially valuable to laboratory researchers and breeders alike. More important, colchicine seemed to alter chromosome number in predictable ways, meaning that it would enable breeders to produce new plant varieties according to a preconceived plan.

If this notion inspired scientists to pursue the production of colchicine-improved varieties, their work in turn inspired rosy visions of a colchicine-improved world. In the national news and specialty magazines alike, colchicine was the "evolution accelerator," a tool that would enable plant breeders "to turn out newly created forms on the production line."[4] And this newfound ability—to turn out plants on the production line—was taken as an indication that plant breeders would play their part, alongside other scientists and industrialists, in producing the goods that would keep Americans strong in wartime and comfortable in the postwar world. Like the x-ray visions of a decade earlier, these expectations for colchicine, articulated in venues from the *Journal of Heredity* to the *New York Times* to *Popular Mechanics*, created a context in which the chemical's uses begged for exploration and exploitation.

For obvious reasons, agricultural plants featured prominently among the early targets of colchicine research. Blakeslee had eagerly reached out to those knowledgeable about plants more valuable than jimson weed, his own intellectually productive but economically irrelevant experimental organism. Meanwhile, Nebel and Ruttle, better positioned to explore the improvement of economic crops from the agricultural experiment station in New York, applied the chemical to fruits and herbs as well as flowers. Around the country, and around the world, other researchers followed suit.

At the University of Missouri, a small USDA project on polyploidy directed by Lewis Stadler had been established before the announcements of 1937. The staff assigned to the program studied genetic problems specific to polyploid crop plants such as wheat and maize. One employee on the project, the geneticist Ernest Sears, had been looking into the question of how to induce chromosome doubling when the colchicine news broke.[5] He immediately began testing the chemical's effects for himself, directing attention toward its potential use in wheat breeding.[6] Sears was particularly keen on doubling chromosomes in a hybrid of wheat and rye—a combination that typically produced sterile offspring. The production of a fertile wheat-rye hybrid had long tantalized breeders, who saw the advantage of combining the nutritional qualities and productivity of the former with the hardiness of the latter.[7]

Researchers tested the capabilities of colchicine on many major cash crops, including wheat, cotton, tobacco, and sugarcane, and on crops of lesser importance, too, anything from berries and brambles to potatoes to pine trees. In 1939, the USDA released a report detailing some of its employees' recent efforts with colchicine: In North Carolina, the geneticist and breeder James Beasley pursued experiments in crossing Sea Island cotton and American upland cotton, hoping to produce a true-breeding fertile hybrid of this sought-after combination through chromosome doubling. In Virginia, Harold Smith of the USDA Experimental Farm attempted colchicine treatment on tobacco hybrids, hoping for similar effects. And at the Bureau of Plant Industry in Beltsville, Maryland, the cytologist Haig Dermen and fruit specialist George Darrow worked out the chemical's effects on a whole range of berries and fruit trees, hoping to produce a fertile hybrid of loganberry and blackberry, a red raspberry suitable for warm climates, improved peaches, apples, cranberries, and more.[8]

Established knowledge of the role of polyploidy in plant evolution, natural and artificial, informed this swift uptake of colchicine. As Dermen of the USDA noted, "the profound interest" in colchicine that could be found "among plant breeders the whole world over" stemmed from their recognition that polyploids were common in nature and frequently possessed desirable characteristics. Dermen also pointed out that colchicine had the advantage of being far more predictable in its effects than any of the older methods for generating polyploids.[9] Researchers had bemoaned the fickleness of methods like exposure to temperature extremes prior to the discovery of colchicine's apparently more reliable outcomes. "Our chief problem seems to be the development of a technique for producing polyploids in considerable numbers," acknowledged Sears in an April 1937 report. At the time, he had several wheat hybrids on which he wished to attempt chromosome doubling but no reliable method to follow.[10] The announcement of colchicine's effects appeared to change those conditions. To Sears and others, it indicated that long-imagined projects like the creation of wheat-rye hybrids were newly viable and could be pursued with greater confidence. In this sense, some breeders did engage in attempts to make varieties "to order."

Others may well have pursued colchicine experiments simply because there was so much discussion of it, in the scientific literature and national news alike. Richard Baker of the University of Nebraska described his study of colchicine effects in potatoes as having been started in late 1938, "during the initial period of enthusiasm for the production of polyploid plants by colchicine treatment." The study was not conducted to produce a specific potato hybrid or desired alteration but rather to determine "the inherent value," if any, "of an increase in the chromosome number of *Solanum tuberosum* [potatoes] for commercial production."[11] For researchers like Baker, inspired by all the talk of colchicine, the aim of experimentation with the chemical was not to carry out a planned breeding project but simply to see what the effects were and to assess whether they might be valuable.

Interest in colchicine quickly grew to such an extent that a few geneticists and breeders grumbled about the research bubble, questioning the immense practical significance that many people had attributed to a single chemical. Reviews of colchicine-related research attest to the growth of the field: 87 references were cited in a review of November 1940; a claim of "more than 200 papers" on induced polyploidy appeared in an article of July 1941; by 1947, a comprehensive bibliography of colchicine contained more than 1,000 references.[12] The Dutch biologist and plant breeder S. J. Wellensiek, observing the proliferation of colchicine research in 1939, referred to it as "the newest fad."[13] G. H. Bates, a breeder at the Farm Institute in Penkridge, England, though in agreement with other breeders that "Mendelian methods have taken us as far as they can," thought that the excitement seen for colchicine by 1939, and especially the claims being made for its revolutionizing agriculture, were likely overblown. "Results up to the moment," he declared, "scarcely seem to justify the description of the new technique as being more than promising."[14]

Some of these complaints were directed more at the fantastical stories that circulated in American newspapers and magazines than at researchers' enthusiasm for colchicine. In their discovery of colchicine's chromosomal effects, and willingness to make big claims for its eventual applications, Blakeslee, Nebel and Ruttle, and others had contributed to the creation of an exciting general interest news story. This was most evident in the pile of newspaper and magazine reports about colchicine-related research that stacked up in the late 1930s and onward. The story line of these reports was almost always the same. Colchicine would, through the efforts of breeders, lead to bigger and better crops. This effect was fundamentally the result of the chemical's ability to "speed up" basic evolutionary processes. And the ultimate payoff would be economic gain, both for farmers and for the nation as a whole.

The first of those elements, the creation of marvelous crops and flowers, was the anticipated outcome of colchicine experimentation most emphasized by journalists writing for a general audience, whether in national or regional newspapers or mass-market magazines. In April 1939, readers of *Country Home Magazine* were regaled with descriptions of giant garden flowers such as zinnias and cosmos, better crops including wheat, tobacco, and alfalfa, and improved peaches and strawberries—all made possible with colchicine. The future looked even rosier, with "blightproof and insectproof vegetables, fruits and forage crops," vegetables able "to fight winter and bloom again in the spring," and "entirely new types of human food from weeds, wild shrubs and forest trees."[15] The journalist Frank Taylor offered a similar assessment, telling readers of *Better Homes and Gardens* to "imagine a day when annuals will be turned into perennials, when new species will multiply and replenish the earth, when plant 'engineers' will build at will varieties never dreamed of by Nature."[16]

Other accounts stressed that this process of enlarging plants was an improvement on evolution itself. The reporter Stephen McDonough told readers about USDA scientists having developed several "new species of plants with the 'evolution accelerator.'" He further explained, "Normally this chromosome-doubling might not happen once in 1000 years to create a new kind of plant. However, with colchicine . . . entirely new species have been created within three years."[17] W. Atlee Burpee & Co. also drew attention for its demonstration of the "evolution accelerator." The Tetra marigold, according to one report, was an example of "a flower actually created by chemically accelerated evolution," evidence of how the "laboratory horticulturist" would enhance the value of crops and the beauty of flowers in years to come.[18]

Colchicine reportedly would make plants grow bigger and evolution run faster; the further implication was that this would make American agriculture more profitable. An article detailing advances in various sciences for the year 1939 singled out colchicine as one of the most promising discoveries of all. "Experimenters believe its possibilities are unlimited and that its use will add hundreds of millions of dollars annually to the Nation's agricultural income," claimed the journalist Warner Oliver.[19] Other accounts agreed. In the estimation of one *New York Times* reporter, colchicine, "the pacemaker of evolution," "a veritable magic wand with which the important plants that feed and clothe man can be commanded to do the impossible," would "add hundreds of millions to the national income."[20]

As such reports indicate, interest in colchicine was tied to expectations for its offering control over evolution and therefore faster and cheaper production of agricultural plants, as had been the case with x-ray-induced mutation a decade earlier. But this was not the only reason why colchicine was of surpassing interest in 1937 and afterward. It also sparked hopes about harnessing evolution to make new life-forms to order in a context in which chemicals were increasingly celebrated as tools that would remake the world, and chemists were the men (and perhaps occasionally the women) who would wield these tools for global benefit.

Even before making tremendous gains as a result of their participation in World War II, the chemical industries—their products and employees—had garnered attention for their contributions to the American economy.[21] Consumer products were in part responsible for this growing visibility, including chemical pharmaceuticals, household goods made from synthetic materials and fabrics, and farm and garden treatments such as artificial fertilizers and insecticides. The development of plastics, and the plethora of affordable consumer goods they made possible, seemed especially to characterize an age of "better living through chemistry."[22] Those working in the chemical industries were particularly eager to highlight such developments, and to suggest that chemistry had reshaped nearly every aspect of modern American life.[23]

Of greatest relevance to the history of colchicine was the influence of the chemical industries on agriculture in the 1930s and 1940s, including the accelerated use of many chemical products such as fertilizers, pesticides, and growth stimulants. American farmers began to use fertilizers at increasing rates in the late 1930s, an uptake explained in part by chemical innovations. Novel industrial processes reduced the costs of producing nitrogen fertilizers, making them much more affordable to farmers.[24] A rise in the use of chemical pesticides occurred shortly thereafter, in the wake of World War II, as chemical producers turned their attention from war needs to potential peacetime profits.[25] The use of herbicides, too, expanded in the 1940s, a development attributed to both the chemical industry and biological researchers whose research on plant hormones revealed the properties that would make these effective agricultural tools.[26]

Many observers—agricultural researchers, popular science writers, and even industrial chemists—considered colchicine a further confirmation of this expanding role for chemicals in agriculture. Not only could these be used to enhance soil fertility, fend off insects, destroy weeds, and speed plant growth; they could now also be used to produce entirely new kinds of crops and flowers. Thinking along just such lines, a contributor to the USDA's 1940 *Yearbook of Agriculture* rounded off a list of chemical advances in agriculture, including fertilizers, pesticides, and hormones, with a mention of the chemical that would "speed up synthetic evolution" by promoting hybridization of types.[27] A journalist for *Popular Mechanics* who interpreted the use of colchicine as a victory of chemical science noted, "Most of this chemical plant growing is in the experimental stage but enough has come from the laboratories to assure us that the era of the chemical plant engineer is here."[28] And a few chemical engineers seem to have been interested in becoming "chemical plant engineers," as their pursuit of information on colchicine suggests. These included Willard Dow, president of Dow Chemical, who expressed interest in learning more about colchicine in the fall of 1937, hoping no doubt to augment his company's investments in products for the farm.[29]

Another indication that ideas about colchicine were entangled in expectations for chemically enhanced agriculture—and home gardening, too, which provided another target market for new chemical products like fertilizers and pesticides—was the frequency with which the chemical was associated with nutrient supplements. The same *Popular Mechanics* journalist mentioned

above described colchicine alongside other "synthetic plant growth substances" that could be applied to improve plant growth in an article entitled
"Growing 'Em Bigger with Chemicals."[30] Similar examples abound: a second
Popular Mechanics article described a range of chemical treatments such as
vitamins and hormones, and included colchicine as "the most promising" of
these treatments; a biology teacher placed it among plant hormones (though
noting it was not a plant hormone) in her 1944 survey of "growth-regulating
substances" for plants; a chemical supplier's advertisement listed "colchisalve"
for home use alongside vitamin B_1 tablets.[31] Gardeners who encountered such
reports could hardly be blamed if they equated colchicine with those treatments already at work in their flowerbeds.

Just as the chemical industries were increasingly insinuated in agricultural
production and home gardening in the 1930s and 1940s, they were also making greater inroads into medicine. Most notably, chemists had participated
in producing new "wonder drugs" such as the antibiotic sulfa drugs released
in the mid-1930s.[32] If there were so-called wonder drugs of the human world,
then colchicine seemed an obvious parallel for plant life. Many journalists
made the connection. "Wonder Plant-Drug Will Help You Grow Giant Vegetables," noted the headline of one piece, which further described colchicine as
a "common drug available through your local druggist."[33] It is easy to see why
"wonder drug" would be a frequent assessment of colchicine. It was, after all,
a compound known for its medicinal properties—most often referred to in
reports as a drug—now being advertised as a route to bigger, stronger, more
vigorous plants. Colchicine could be applied with an eyedropper, rubbed on as
a salve, or injected directly into stalks and roots. Even the hypodermic syringe,
used for delivering an antibiotic or a vaccine in the doctor's office, could now
be used in the farm or garden (fig. 16). "'Shots in arm' for plants are used to
improve breed," declared one article.[34] A few described the use of colchicine
as "doping" plants, a process that "tricked" nature into producing more at a
faster pace.[35] Such reports suggested that the chemical could be understood as
having an unnatural, though not necessarily undesirable, effect. Rather than
cure plants of some ailment, or improve their overall health and vigor, it would
push them beyond natural limits.

Clearly the welcome given to the chemical "evolution accelerator" was
augmented by a wider celebratory attitude toward chemicals. Chemical treatments had enabled greater control over weeds and pests in agricultural fields
and home gardens, and over bacteria and other disease-causing agents in the
hospital—so why not over genes and chromosomes as well? Just as there was
"better living through chemistry," so too did it seem that there could be better
evolution through chemistry.

FIG. 3.—Growing tip of a bulb treated with colchicine solution by means of a hypodermic needle.

FIGURE 16. A textbook illustration depicts the injection of colchicine into a plant bulb. From G. S. Avery, *Hormones in Horticulture: The Use of Special Chemicals in the Control of Plant Growth* (New York: McGraw-Hill, 1947), 287.

Were this not motivation enough, the outbreak of war provided still further reasons to promote colchicine experimentation. The United States entered World War II in December 1941 and remained at war until August 1945. In wartime, it seemed necessary to push many things beyond their known limits—not just plants, or evolution, but also people and industries. Colchicine breeding, much like other advances in science and technology, would help make the nation as a whole more self-sufficient. For example, some researchers hoped that colchicine could alleviate rubber shortages precipitated by Japan's conquest of the South Pacific in 1942.[36] Colchicine offered a promising lead to greater domestic production: perhaps it would be possible to increase the yield of particular rubber-producing plants through induced tetraploidy. At Cold Spring Harbor, two researchers took up work along these lines in 1942 after the USDA Bureau of Plant Industry solicited their aid.[37]

Colchicine might aid American self-sufficiency in still other ways. The expectation that all Americans would participate in the war effort, whether in their commitment to increased industrial or agricultural production or in their efforts to produce and conserve within their homes, dominated on the home front.[38] One of the most successful volunteer campaigns of the war was the call for families to establish "victory gardens." The National Victory Gardening Program, begun after the attack on Pearl Harbor, encouraged families to become more self-sufficient in food production so that crops grown on American farms could be sent to support the war overseas without citizens at home feeling the pinch. By 1943, some three-fifths of Americans were cultivating gardens; their efforts that year accounted for about 40 percent of the fresh produce consumed by Americans.[39] It stands to reason that a treatment promising to produce bigger and better vegetables would be well received at such a time. Some predicted that scientists and breeders would use colchicine and other techniques to improve products for home growers. "Victory gardens next year may increase production," noted a science journalist in 1942, "as plant breeders develop improved varieties . . . and experiment with hormones and other strange chemicals such as colchicine, which so change the plant's heredity as sometimes to produce a super-plant."[40] Others proposed home experimentation with colchicine as yet another way for civilians to contribute to the war effort alongside buying war bonds or taking a job in industrial production. A display at the December 1941 AAAS meeting hailed "war gardens of science" in which new scientific techniques would be applied to make victory gardens "sources of more, better and less costly foods and vitamins." Colchicine numbered among the tools that would make these improvements possible.[41] Sister Mary de Lourdes, a biology teacher, recommended that with "Victory Gardens now in vogue," teaching high school students how to use colchicine would be an ideal route for introducing its use in home and community gardens.[42] At least one school did just that, including lessons in "chang[ing] certain hereditary factors of a plant" with colchicine among various projects meant to increase food production and encourage wise consumption.[43]

In some accounts, colchicine would also help rebuild the postwar world. A *Newsweek* article of 1944 regaled readers with descriptions of what agriculture, unhindered by the heavy demand on resources required by war and newly aided by genetic science, would look like. The genetic experiments of the 1940s would lead to "myriad unfamiliar fruits, all painstakingly designed . . . plants nature alone could never have bred." These would be novel in taste, color, and texture, and would of course be of tremendous size. All of this would be the handiwork of "geneticists and plant breeders" who were "patiently molding the germ plasm of green things to man's needs." *Newsweek* raised some concerns about what these would mean for the future. Plants bred to be machine picked and highly prolific could mean unemployment for farm workers or overproduction of crops unless new channels for distribution could be created. Still these risks were thought to be worth taking, because science promised to regreen a ravaged planet. As the report concluded, "The seeds now germinating can remain in their clay pots and green-house trays as laboratory curiosities, or they can multiply and replenish the worn earth."[44]

Replenishment was obviously the preferred route, and one that meshed well with the direction of agricultural production in the United States and many places around the world after the war had ended. World War II marked a significant turning point in US agriculture. As a result of the introduction of higher-yielding breeds of many economic crops, the mechanization of labor, and increased chemical inputs, productivity shot upward and prices fell. American consumers in the years after the war found that they had access to a greater variety of food products at a lower cost than ever before.[45] Meanwhile, the Rockefeller Foundation was supporting agricultural research and fieldwork that would eventually form the basis of what we now call the Green Revolution. This surge in agricultural productivity in a number of developing nations in the 1950s and 1960s, which resulted from the introduction of new breeds of crops (produced via well-established methods such as selection and hybridization) and the increased use of synthetic fertilizers and pesticides along with other adaptations of developed-world agricultural methods and economic supports, would eventually bring farmers around the world into the style and tempo of production that had come to dominate American farming.[46] In this arena, too, of modernizing (or "modernizing") agriculture to alleviate hunger and streamline agricultural economies, colchicine could be slotted into the general pattern of agricultural progress. With many more mouths to feed, a tool that promised better breeds at an increased pace could only be a boon. "In an ever-hungry world, nature needs to be given a hot-foot," wrote one journalist. "Evolution must be given better pickup, quicker getaway."[47] Otherwise, the constant pressure of a growing human population would overtake nature's inherent limitations.

Whether speaking of speed and size, efficiency and productivity, or global needs and modernization, colchicine seemed to fit the bill drawn up for food and agricultural production, both in wartime and in the postwar world. This explains, in part, why experimentation with colchicine held such fascination. The chemical would allow geneticists and breeders to take their place among the many scientists and technologists contributing to national growth and development by facilitating the production of new and improved agricultural varieties through a seemingly precise and predictable technique. Like x-ray-induced mutation, the draw of colchicine-induced chromosome manipulation was for many Americans closely linked to the growth of industry and the hope for more efficient agricultural production—only this time through the use of a chemical technology.

This is only part of the story, however. If the central appeal of colchicine to many professional plant scientists and other observers lay in the apparent advantage of calculable genetic alteration and efficient agricultural production via a reliable chemical process, other constituencies clamored for colchicine with very different purposes in mind.

Tinkering Technologists

Part of the mystique of colchicine to Americans of the 1940s was that it seemed so easy to use. Reports described how scientists produced the desired effects by "simply spraying the plants, or dipping them into a solution of the chemical."[1] They referred to colchicine as a "short-cut" that would avoid the long and painstaking processes associated with older methods of plant breeding like selection and hybridization.[2] In fact, they suggested that the application of colchicine was so straightforward that even the amateur experimenter could manage it effectively. "If you have ever done any gardening, you might have an urge to try your own hand at changing chromosomes. There is no secret about it," encouraged one reporter in 1939. "In contrast to many scientific experiments, this one is easy to perform."[3]

Exaggerated reports of spray-on plant improvement were intended to engage readers of newspapers and magazines. But an additional consequence of such reports, likely unintended, was the pursuit of colchicine research by enthusiastic amateurs.[4] In their independent explorations, these experimenters call to mind other, better-known, amateur experimenters and tinkerers of the twentieth century, whether ham radio operators, rocketeers, or television and computer hobbyists. As historians have described, these various tinkering technologists, who sought to master technical skills for their own recreation, contributed to the development of certain technologies—and the creation of markets for these technologies as well. They established hobbyist communities, of varying degrees of organization, centered on the circulation of tools and information. They engaged with professionals working in the same domains but were not limited to the ideas and methods those professionals espoused.[5]

Similar activities can be seen among amateurs who experimented with colchicine. They hoped to take this cutting-edge tool of genetics and breeding and direct its use to their own purposes. They sometimes aimed to produce an improved plant, but in other cases wanted simply to see the chemical's effects for themselves. They sought information from professionals and consulted publications directed at amateurs. Some shared their efforts with fellow experimenters, in newspaper articles, garden magazines, and flower societies. And their interest created markets for colchicine that would not otherwise have existed, such as preparations specifically designed for use in home experiments. The extent to which amateurs extended the use of colchicine beyond what most professionals had ever envisioned is best illustrated by the efforts of some professionals to discredit the idea that plant breeding with colchicine could be a leisure activity. Still, despite these warnings and the actual challenges involved in producing something useful, many amateurs tinkered on. In doing so, they created a space in which colchicine did not need to be a technology for precision breeding or genetics engineering but simply a tool to create unusual plants—good or bad—with little effort and for pure leisure. Just tinkering, whether with the technique or with genes and chromosomes, was ambition enough.

Blakeslee and Avery may well have touched off interest in casual experimentation with colchicine through their initial article in 1937. There they suggested that "juggling chromosomes for the betterment of plant-kind is primarily a matter for the trained genetics engineer who knows the chromosomes with which he is working." But the pair then went on to explain how one could obtain colchicine and what it would cost, as well as how to identify a tetraploid by examining pollen under a microscope, a task with which "any high school biology teacher" could assist.[6] Subsequent reports took this idea and ran with it, offering accounts of colchicine research and, in some cases, simple instructions for its application. The science journalist Frank Thone emphasized the quick results of colchicine as a particular selling point for the amateur. "It is even possible for the ordinary gardener to enjoy the thrill of creation, by producing a new . . . flower or vegetable variety," he declared, pointing out that it would be faster and easier than older, more "tedious" methods.[7]

If one were to take these reports at face value, it might seem that the hardest part of conducting an experiment with colchicine would be obtaining a preparation of the chemical. Even this task was straightforward, however. In fact, it was trivial compared with gaining access to x-ray apparatus, the key challenge that faced those amateur breeders who had aspired to x-ray breeding (and there had been a few).[8] The use of colchicine as a therapeutic for gout meant that it could be ordered from a chemical supplier or even purchased from a pharmacist. It did not come cheap: in 1939, a preparation of colchicine from a chemical supplier cost about $25 per ounce (the equivalent of about $415 today).[9] But reports were also quick to mention that very little of the chemical was required for a successful treatment.

From 1937 onward, many farmers, gardeners, and other curious readers—likely inspired by news reports—wrote to professional scientists, seeking advice on how best to conduct their own experiments with colchicine. Orie Eigsti, by that time an assistant professor at the University of Oklahoma, received correspondence from some 1,100 people in response to publicity surrounding a collaborative project with colchicine begun in 1940.[10] Blakeslee reported having received a similar number of letters by 1940, most from outside the world of professional genetics and plant breeding.[11] To deal with the influx of inquires, Blakeslee drew up a form letter that contained his responses to the most common inquiries from those who were clearly amateur: no, colchicine was not a miracle drug for growth enhancement; yes, it was easily obtained and relatively easy to apply (he included instructions on both); no, one could not make money by growing the autumn crocus as a source of colchicine; no, the chemical could not be applied to animals; and, please, you should consult your nearest agricultural experiment station before investing much time or money in colchicine breeding.[12] Blakeslee once described the motley assemblage that wrote to him in search of instruction as "all kinds of people: doctors, druggists, farmers, auto mechanics, housewives—in addition to experts in plant breeding."[13] He might have added that it was an international group, and that it included young people (students in search of school projects especially) in addition to adults.[14]

Many who wrote to Blakeslee and other researchers for assistance were interested in improving crops and flowers. Those writing from the farm often hoped for advice on agricultural plants. One self-described "ag college graduate" expressed interest in breeding better forage sorghums with colchicine.[15] Another recent college graduate, a Mississippi man about to return to work on his father's plantation, sought advice on colchicine to improve their farm output. "I would like to have a little more knowledge about it instead of stumbling blindly along," he requested, and luckily, too, since he hoped to use the chemical on poultry and livestock in addition to crops.[16] A "young farmer of 20" wanted advice for using colchicine on a whole range of plants: corn, beans, oats, wheat, alfalfa, and clover seeds.[17] A Florida man, disappointed that his local agricultural experiment station was not doing any research on colchicine-improved mangoes, thought he would take on this research himself.[18] Still other letters came from home gardeners, who often hoped to produce flower novelties. Many self-described hobbyists asked about using colchicine in their gardening projects, whether breeding new types of iris flowers, producing doubled-sized orchids, or "pepping up" cactus plants.[19] Some correspondents wanted to improve the outcome of the annual harvest from their backyard vegetable patch. "It is my hobby to raise a vegetable garden, and a large one," explained one gardener to Blakeslee and Avery, before asking for advice on whether colchicine would work on all vegetables.[20]

Some correspondents emphasized their interest in experimentation. "I am a very keenly interested experimenter in Iris Breeding," explained one Colorado man in a letter to Blakeslee and Avery.[21] Another, also a Colorado resident, described herself as "an amateur—a garden hobbyist, but deeply interested in improvements in plant life" who "would like to use Colchicine in some experiments."[22] A man seeking information on both "plant growth substances" and colchicine similarly declared, "I am very much interested in experimenting with plant life."[23] In fact, some letter writers seemed more keen on experimentation itself as a hobby. One such letter to Blakeslee arrived from the Delaware County Prison in Thorton, Pennsylvania, where an inmate asked whether anyone had yet tested colchicine on perennial bulbs. Lest Blakeslee think the process too complicated or laborious, the inmate pointed out, "I have the facilities and the 'time' to conduct field work."[24]

Demand for colchicine did appear to increase in light of the attention the chemical attracted from these diverse quarters after 1937. One 1939 report claimed that a chemical wholesaler was selling fifty times the quantity of colchicine he had sold prior to the discovery of colchicine-induced polyploidy.[25] Frank Thone reported in 1941 that demand for colchicine had gotten to be so great that even pharmacists who had not carried it in the past were making a point of adding it to their stock.[26] Soon colchicine was packaged specifically for amateur experiments (fig. 17). One enterprising firm, Cambridge Laboratories, of Wellesley, Massachusetts, in 1940 offered colchicine for sale by mail order in a variety of quantities and preparations, ranging from $21 for one-half ounce, to $2 for either a 100-milliliter vial of 0.6 percent solution or a small vial of "colchisalve." Even if the reader had not yet heard of colchicine, he or she might be tantalized by the company's claim that the "easy to use drug . . . doubles, quadruples chromosomes, produces giant never before existing plants, huge fruits, doubled flowers."[27]

Once one had obtained colchicine, instructions on how to use it could come from a variety of sources: response letters from professionals like Eigsti and Blakeslee, articles appearing in science magazines like *Popular Mechanics*, or newspaper stories about colchicine. Even the *Journal of Heredity* participated in the provision of instructions for would-be experimenters. In response to an influx of inquiries on how to use colchicine, the journal assembled a list of its publications on the subject. It offered these for sale as reprints at

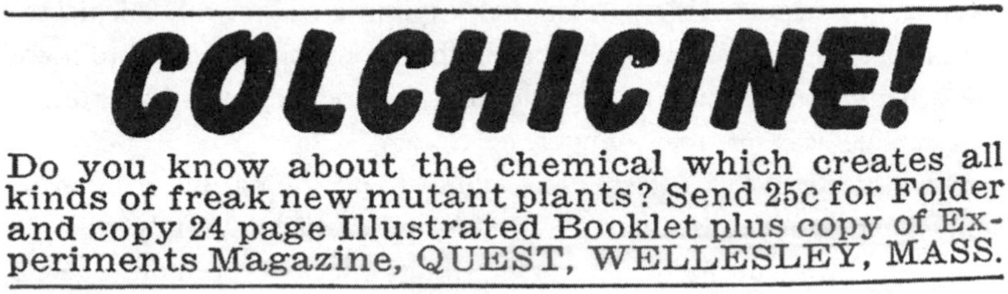

COLCHICINE!

Do you know about the chemical which creates all kinds of freak new mutant plants? Send 25c for Folder and copy 24 page Illustrated Booklet plus copy of Experiments Magazine, QUEST, WELLESLEY, MASS.

FIGURE 17. This advertisement for a commercial preparation of colchicine makes a pitch to gardeners who might want to tinker with their backyard varieties. From *Popular Mechanics*, April 1943, 24A.

thirty-five cents per copy, except for Blakeslee and Avery's article of December 1937, which sold for one dollar. The journal clearly targeted a range of experimenters with these sales. The editor emphasized that the cytological methods described in many of the articles—including preparing slides of cell nuclei to see whether the chromosome number had in fact changed—were not absolutely necessary, and that "the amateur breeder . . . may still get valuable results merely by using his unaided eyes."[28]

Articles and book chapters about colchicine published for amateurs typically assured readers that those who followed appropriate methods could expect good results, as judged by the production of calculated improvements in their flowers and vegetables. A chapter on the chemical "miracle worker" (i.e., colchicine) in a 1941 garden manual, *Science in the Garden*, reported that following the instructions contained therein could result in "interesting, beautiful, and economically important" varieties. The authors suggested that the amateur gardener-scientist set to work solving problems of hybrid sterility or failed crosses due to differences in chromosome numbers, providing a sample experiment for the latter in which two incompatible varieties of phlox could be made crossable.[29]

Some amateurs actually applied their newfound knowledge of colchicine to their flower and vegetable gardens, many of them hoping to produce an improved plant. A Seattle man, who "for many years, as a hobby," had been "hybridizing carrots, in an effort to get a pure golden yellow color," tested colchicine to see if it would help him achieve this goal. (It didn't.)[30] Another diligent amateur flower cultivator described his experiments on lupines, shasta daisies, and gladiolus flowers to the magazine *Horticulture* in 1940. His work had not produced anything of value—nearly every one of his plants had died—yet he thought "the difficulties encountered may be of interest" to other readers, so he offered them for general circulation.[31] His report prompted a response from a representative of the Empire State Gladiolus Society, who wrote to affirm the society's determination to "lend its support in a wholehearted way to the amateur breeders of the state" in their experimentation with colchicine.[32] Some experimenters provided reports on their colchicine research to flower and garden societies, where an altered garden rhododendron flower or indoor African violet was sure to be of interest.[33]

Other amateurs focused less on developing improved plants or following planned experiments, and more on producing novelties or simply satisfying their curiosity about the "magic spray" and its effects. W. E. Bott, a realtor from suburban Ohio with an amateur passion for "horticultural chemistry," counted among this group. He carried out his own experiments beginning in 1938, following up on some of the reports circulating in the press. "I have been a little indiscriminate in my use of colchicine," he admitted, having tried it on nearly everything growing in his garden. The results of this exuberant application had proved thrilling, his plants growing weirdly and wildly. His marigolds, columbines, and four o'clocks were "behaving as if they were bewitched." His snowball bush had blossoms "as big as a dinner plate." One zinnia was somersaulting about his small garden in corkscrew loops, a trait he speculated had never before been seen in this species. Bott felt assured of a future world "filled with super-flowers, super-vegetables, super-trees and plants of every kind" and eagerly shared his methods in a three-part series in the local newspaper so that other gardeners might follow his lead. He emphasized above all that amateur experimenters need not justify their actions in terms of benefits to agriculture or horticulture: "It is true that some amateurs . . . may be able to make important discoveries. Others may contribute much to the knowledge which today's research horticulturists so eagerly seek. . . . Or, if these goals are too pretentious, there is that of pure entertainment—the pleasure of working for horticultural miracles with a good chance of seeing them take place."[34] The notion of pursuing colchicine experiments out of curiosity and adventure, to see what might happen as much as to improve plants, would become a defining feature of amateur work with colchicine in subsequent years.[35]

This growing curiosity about colchicine—and desire to experiment with it—provoked a mixed response from professionals. A few encouraged such activities. Eigsti at Oklahoma began a collaborative project with home gardeners and other amateur experimenters in the spring of 1940. After a brief news item emphasized the cramped conditions under which he was forced to carry out his colchicine research at the University of Oklahoma, Eigsti reportedly was inundated with letters from farmers and gardeners who wanted to provide him with space and assistance.[36] He decided to take on these volunteers as research assistants, offering vials of colchicine solution to 100 amateurs who would track the progress of their experiments and send him the results.[37] Eigsti received far more responses than he had anticipated and was quickly out of the solution. The following year, he sought and received funding from the Carnegie Corporation of New York to expand the project. He succeeded in enlisting 327 volunteers from 38 states and several foreign countries.[38]

The "direct objective" of the project, as detailed in the final report, was to gather information about how "the layman can use the colchicine method" and to teach him that "the colchicine treatment is but one step in a program of plant improvement."[39] Not that the volunteers were to consider themselves either test subjects or students, as that description implied: the materials circulated to participants in the project—called the "Amateur Co-operative Research Project"—encouraged the volunteers to think of themselves as scientists whose results would generate new knowledge. Along with the vial of colchicine, participants received detailed instructions on what types of plants to select for the experiment, how to determine an appropriate method of treating their particular test plants, how to keep controls, what to look for in treated plants, and so on.[40] The program also encouraged amateurs to think that they had a reasonable chance of producing something worthwhile. Participants had to sign an agreement with the University of Oklahoma, which ensured that if any participant produced a valuable plant the university would see some of the profit.[41]

Thanks mostly to the eager news reporting of the Science Service journalist Frank Thone, Eigsti's project received considerable media attention. Thone portrayed the work as an invitation to amateurs to contribute both to agriculture and to scientific knowledge. "You don't have to be a Ph.D. in botany to make valuable contributions in the field of plant breeding," he wrote in a 1941 article that described Eigsti and his "far-flung army of scientific collaborators"—an article that also included instructions on using colchicine for those who had not been lucky enough to be involved.[42] Those who were skeptical of such a claim, or who doubted that Eigsti's untrained volunteers had actually produced anything of interest, need only consider the results obtained by some participants. An Oklahoma schoolteacher named Leona Schnell reportedly produced an extra-large periwinkle, "superior in size and beauty and resistance to drought and diseases," from her experiments. And Eunice Moore, a nurse from Tulsa, Oklahoma, had used the colchicine solution to produce soybeans twice as big as usual and containing double the typical amount of oil. According to one account, it had required very little of Moore to achieve these economically promising soybeans. "She simply followed Dr. Eigsti's directions" and—presto!—the monster-sized beans emerged.[43]

Other scientists were more wary of getting nonprofessionals involved in experimentation. One concern was that colchicine was toxic and inexperienced experimenters might do themselves harm. The *Journal of Heredity*, though it seemed content to circulate information about colchicine research to anyone who would purchase it, also warned would-be experimenters of potential hazards in an editorial published after Blakeslee and Avery's 1937 article.[44] This warning appeared again with more substantive illustration the following year because, the editor noted, "colchicine is being so widely used." The second notice cautioned that even dilute solutions of the chemical would be damaging to skin if not washed immediately, and included a personal account from the editor, who had accidentally gotten colchicine dust in his eye. Although the resulting inflammation and blindness had proved temporary, "and the eye does not seem to be a tetraploid at this writing," the editor hoped his tale would prevent others from repeating his "involuntary experiment."[45] Haig Dermen of the USDA attempted another public warning in 1940, emphasizing once again that the chemical needed to be handled with care. His warning was unlikely to have been very dissuasive, at least as it appeared in one printing. "Colchicine, the 'magic drug' that speeds up evolutionary processes in plants . . . may be dangerous if not handled with proper precautions," Dermen was reported to have warned the "hundreds of enthusiastic amateur experimenters now engaged in trying to produce strange and possibly valuable new varieties."[46] The possibility of bodily harm might well have seemed a small risk to take for such rewards.

A few professionals thought that warnings did not go far enough—that the public ought to know that colchicine was not the miracle worker that most articles claimed it to be. Gordon Morrison, a plant breeder at the Ferry-Morse seed company, thought he might bring the sky-high expectations of gardeners back down to earth by providing readers of *Gardeners' Chronicle* with only the "Facts about Colchicine." Morrison pointed out that although many plants had successfully had their chromosome numbers doubled, and although a few of these instances produced superior types, colchicine was "not a biological 'Aladdin's Lamp'" for producing giant plants. Some species would be unresponsive to the treatment, and others had mechanical limits for size that no chemical stimulant or chromosome manipulation was likely to overcome.[47] Gardening columnists and other critics voiced similar concerns, complaining about fantastical claims of "family-sized vegetables" and "Shasta daisies as large as dinner plates."[48] They fumed at the public being fed what they perceived to be misleading information. But those who groused about or even attempted to dampen colchicine enthusiasm in the late 1930s and early 1940s fought a losing battle, for celebratory—and exaggeratory—articles rolled off the presses for years. Their titles suggest their key content: "Agriculture's Giant Powder" (*Country Home Magazine*, 1939), "Giants in the Garden" (*American*

Burpee worker sprays young snapdragons with colchicine solution to create outstanding new varieties.

Here's a 32-pound turnip that dwarfs a 3-year-old
child. Will all turnips in the future be this size?

FIGURE 18. "Will all turnips in the future be this size?" asks a magazine article that surely made a little
too much of colchicine research. In another frame, an employee of W. Atlee Burpee & Co. demonstrates the
quick and easy nature of plant breeding with colchicine: Simply spray young plants "to create outstanding
new varieties"! From Donald G. Cooley, "Apples as Big as Your Head!" *Mechanix Illustrated*, March 1949, 81.

Magazine, 1940), "Remaking the World of Plants" (*New Republic*, 1941), "Plant
Magicians Now at Work Improving World's Food Crops" (*Newsweek*, 1944),
"New Plant World A'Coming!" (*Coronet*, 1945).[49]

Each of these articles in turn prompted a wave of letters from readers
who wanted instructions on how to use colchicine. Blakeslee had complained
about this phenomenon early on, assessing the intelligence of the letters by the
comparative luridness of the inspiring article.[50] The phenomenon continued.
A 1949 article in *Mechanix Illustrated*, appearing nearly twelve years after the
first announcements of colchicine-induced polyploidy and entitled "Apples as
Big as Your Head!" described the usual oversized fruits, vegetables, and flowers that colchicine could be expected to produce and included photographs
to illustrate: a three-year-old child standing next to a turnip nearly as tall as
she is, and a woman's head peeking out from behind an enormous "king-size"
cabbage (fig. 18). For those who were not interested in extra-large veggies, the
author suggested that an "exotic new fruit or vegetable" would be more appealing: the "pomato" or the "cucaloupe," for example.[51] Predictably, he attributed
these developments to scientists' "new" ability to speed up the ordinarily slow

process of evolution. And he encouraged his readers to try it for themselves, providing instructions and promising fast results. Researchers soon reaped the true harvest from such reporting, a crop of misinformed amateurs. "I have read in the March Mechanix Illustrated, concerning plants that can be grown to enormous sizes by a Drug called Colchicine," wrote one inquiring farmer to Blakeslee.[52] Writing in March 1949, presumably in response to the same article and sounding like a sensational news writer herself, Mrs. H. M. Turner asked for information and literature regarding "'colchicine' a yellowish white powder which has caused a veritable revolution in the plant world," as well as for a confirmation that the eating qualities of larger fruits and vegetables would remain just as good as the prerevolution types.[53]

Still, even if the majority of newspaper and magazine reports about colchicine had been more restrained in their claims about what could be created with the chemical, as some professional researchers had wished, it is not at all clear that this would have eliminated amateur interest in colchicine. As described above, not every amateur was interested solely, or even primarily, in improved flowers and vegetables. Experimentation itself had some appeal. Just as early radio hobbyists and later computer tinkerers found enjoyment in the very act of tinkering with these mechanical devices, so too could an amateur breeder or experimenter find enjoyment in tinkering with the internal mechanisms of garden flowers or farm crops, regardless of the specific form of the final product.[54]

As late as 1961, magazines continued to tout colchicine as an ideal experiment for an amateur interested in altering his or her plants, and not just to produce bigger flowers or new hybrids. Readers of *Popular Mechanics* were encouraged to think of colchicine as suitable for "the experimentations of serious amateur gardeners," and they received instructions on how to carry out such experiments. A few things had changed in two decades, however. For one, a little background in science was thought to be helpful. The author counseled his readers to "read a book or two to learn something about genetics." And perhaps informed by just how slow achievements with colchicine were in coming, even among professionals, he did not promise amateurs the "spectacular results" that professionals and commercial developers had experienced but something rather more modest: an effect. "Good or bad, you'll get *changes in plants*," he wrote, as if to suggest that this alone constituted success.[55]

The early, and continuing, interest in colchicine among amateurs was not linked exclusively to the precise production of new varieties—the vision that inspired many professional breeders. It also emerged from a more general interest in tinkering with genes and chromosomes, in playing around with colchicine and plant seeds just to see what might result. The Ohio amateur

Bott summed it up best in the account of his colchicine experiments he provided to fellow gardeners. "The production of botanical 'accidents' or 'sports' has the same fascination for me that it should have for amateur gardeners everywhere," he declared. Colchicine would be sure to provide these, if nothing else. Bott characterized the likely results: "Not all of them [the experiments] will be worthwhile, of course. Perhaps even the great majority will produce plants that no one would want to perpetuate." This was not an issue. "The important thing is that the mutations can be caused by methods so simple that any amateur can apply them." Random mutations, not improved plants, were Bott's central goal. Knowing that the chemical was altering the fundamental components of heredity would serve only "to intensify the incredulity you feel at witnessing colchicine's power."[56]

This curiosity about "witnessing colchicine's power"—that is, tinkering with colchicine, and with genes and chromosomes, simply to see what might result—was the key attraction. And, as it turns out, this particular goal was not exclusive to amateurs like Bott.

The Flower Manufacturers

Amateur experimenters found themselves in good company in their efforts to turn out novelties and curiosities for the garden. Both sellers and growers in the flower-seed market placed a premium on novelty, whether this came in the form of a new or unusual color, double flowers, or variegated leaves—any characteristic that might distinguish an ordinary plant from an extraordinary one. This "insatiable appetite for novelty," as one breeder called it, caused some observers to speculate that there were likely to be more opportunities for the application of colchicine in creating garden flowers than in food plants.[1] These predictions came to pass, and there is no better evidence than the part played by W. Atlee Burpee & Co. (hereafter, Burpee Seed) and its marigold-loving president, David Burpee, in the larger colchicine fad. Burpee Seed was the first company to introduce to the market a colchicine-created product, the Tetra marigold of 1940, and it followed up with a whole range of chromosome-doubled flowers in the next couple of decades. These various, and vigorous, tetraploid flowers were proof that colchicine could be a useful tool for generating garden novelties.

The importance of novelty extended beyond the physical forms of the tetraploid types. Burpee Seed, and especially David Burpee himself, hawked these as products of the very latest science and technology. The novelty of their creation was as much a selling point as their unusual blooms or size or vigor. For this reason, colchicine was not always set apart by Burpee as a unique technology that enabled planned production of new varieties, but rather was placed among a whole array of "scientifically advanced" methods that allowed him and his breeders to both generate novelties and advertise their novel production. X-rays, chemicals, radioisotopes—Burpee breeders wielded these at various experimental farms, trying to generate plants, while

David Burpee touted them in interviews with journalists and presentations to garden societies, trying to generate customers. And Burpee breeders often (though not always) used these various tools in similar ways, as mutagens that might produce a random change.

David Burpee's use of colchicine speaks to the continued importance of tinkering in plant innovation, even as he and others increasingly emphasized streamlined engineering or manufacturing approaches. More important, it highlights how tinkering to produce biological innovations—like new flower novelties—could apply to both work with the techniques and tools and work with genes and chromosomes. Just as researchers had long tinkered with experimental methods of inducing chromosome doubling, Burpee breeders tinkered with the methods of colchicine application to find out what approaches would work best on their various crops and flowers. But that was not the sole extent of their tinkering. They also used colchicine to tinker with the hereditary material of plants, trying to produce new arrangements and combinations and therefore novel varieties. As I argue here, drawing especially on the work carried out at Burpee Seed, although some geneticists and breeders had seen colchicine as a way to escape their reliance on tinkering in both the innovation of breeding methods and the innovation of plant varieties, colchicine's more immediate (and to some extent enduring) appeal turned out to be in enabling and even expanding these selfsame tinkering activities.

This realization in turn drives home another point, one familiar to historians of technology. When it comes to understanding the history of this particular genetic technology, it is essential to differentiate among its varied users.[2] Colchicine, as we will see, fell far short of the ambitious claims made for it as a tool of genetics engineering or precision breeding. But not everyone was interested in making plants "to order." It could be equally interesting and useful just to produce changes, which is what some users—even professional users—wanted. And with colchicine, it was almost always possible to do at least that.

A young David Burpee formally entered the flower-and-vegetable seed industry in 1915, dropping out of Cornell University to take over the family business after the death of his father, W. Atlee Burpee. The twenty-two-year-old Burpee proved more than equal to the task of running Burpee Seed, which at the time was one of the largest mail-order seed companies in the world, sending out some one million catalogs per year. During his nearly six decades at the helm of the company, the charismatic Burpee gained notoriety for his creative and relentless promotion of his company's products and his vigorous pursuit of new and better plant varieties—and for his passion for marigold flowers above all other kinds.[3] By the 1930s, David Burpee was arguably the country's best-known horticulturist.

As America's iconic flower breeder, Burpee took on the legacy of the once-revered Luther Burbank. The connection between the two men was direct: in September 1931, five years after the death of Burbank, Burpee purchased the seed and bulb portion of the Burbank estate from Burbank's widow, Elizabeth. Journalists commenting on the sale equated this physical transfer of seed with the less tangible transfer of power and expertise. As *Time* magazine reported that year, "It has long been the ambition of David Burpee to take over the unfinished work of Luther Burbank. And that work he will carry on, he says, as Burbank did: in a scientific spirit, not a commercial one, in the interest of mankind."[4]

Other observers were quick to note that in fact much had changed within this single generation. They saw Burpee as more scientifically advanced, more cutting-edge—perhaps in part because this is how Burpee liked to see himself. As the journalist Frank Taylor described, drawing on interviews with Burpee and his employees, researchers at seed companies such as Burpee Seed were "typical of the Luther Burbanks who cook up your next year's flowers." Except, however, they could achieve even more, "perform[ing] tricks of horticultural wizardry never dreamed of by Burbank."[5] Burbank's work had been limited to crossing and selection, then the mainstays of practical breeding. He had sought out plants with desirable traits and used selection and hybridization to combine these traits into individual plants and amplify them over time. Burpee, with access to mutation-inducing chemicals and other laboratory methods, sought to create in plants the traits he and his customers desired. Where Burbank had plodded along with these old techniques, Burpee's methods were more advanced, and flashier, too.

As a result of Burpee's interest in novel techniques of plant breeding, the first plant to be introduced on the market as a colchicine-created tetraploid was not a new wheat variety, a large and juicy apple, a taller timber tree, or some other improved food or crop plant. It was the extravagantly large "Giant Tetra Marigold"—with "Tetra" being shorthand for tetraploid. Advertisements offered a size comparison of the tetraploid flower, a "man-made miracle," versus its diploid forebears so that potential buyers could marvel at the difference between them. Descriptions of the flower were equally tantalizing. Readers of *Popular Science* learned not only of its large size and vibrant color, but that it was "said to stay fresh, when cut and brought indoors, far longer than any other species."[6] Of course, readers of the magazine would have also learned to associate the flower, and its appealing properties, with the new process of "chemically accelerated evolution." But where had this "man-made miracle" come from? Although Burpee described his company's California flower plantation "Floradale" and its accompanying laboratories as places where "varieties are being continually improved and new ones created," the Tetra marigold, in all likelihood, had not been innovated at the Floradale labs.[7] The Guinea Gold marigold, from which the Tetra had been produced, had been among the earliest test subjects of Nebel and Ruttle in New York in their work with colchicine. It was only after learning of Nebel and Ruttle's research that Burpee had set the breeders under his employ to work using the chemical.

In short order, Burpee breeders would count among the most visible users of colchicine methods. In the subsequent two decades, the company offered for sale tetraploid varieties of a number of popular garden plants, including marigolds, phlox, snapdragons, and zinnias. All were described along the same lines as the Tetra marigold had been: larger, stronger, and lusher than their diploid progenitors, and all products of a chromosome-doubling compound called colchicine. The snapdragons seemed to benefit the most from the chemical enhancement. By the mid-1950s, Burpee's Tetra Snaps were the best-selling varieties of the plant in the United States.[8]

Colchicine provided David Burpee with an obvious route to advertise his company's place on the cutting edge of technological development. His predecessors had relied on the appearance of mutations, or "sports," as the source of new colors, improved shapes, and other desirable characteristics. Burpee believed he had a decisive advantage over these breeders: "In my father's time a 'break,' as the plant breeder usually calls a sport, was supposed to occur once in every 900,000 plants. But now, by artificial stimulants, we can turn them out once in every 900 plants. Or oftener." With colchicine, he saw himself speeding up the process of plant development beyond anything previously considered possible. "Shocking plants to juggle their chromosome arrangement . . . opens a brand new vista," Burpee reported. Of course, it was not only colchicine that opened such vistas. Burpee celebrated other mutation techniques that similarly promised to make breeding faster, more efficient, and more scientific. These included exposure to x-rays and other types of radiation, treatment with chemicals, and deliberate mutilation.[9] One of Burpee's employees, the breeder William Hoag, reportedly described it another way: "We're not just picking up sports, we're manufacturing new flowers scientifically."[10]

And yet "manufacturing new flowers scientifically," if that was meant to convey the idea of planned and predictable improvement, was hardly an appropriate description of what the Burpee breeders were up to with their mutation regimes. For the most part, they were seeking to produce random changes, in the hope that some unexpected genetic mutation or other novel inheritable alteration might prove useful—or, more precisely, sellable. Even the application of colchicine, a technique touted by other breeders as allowing the deliberate design of new types, seems often to have been used (and celebrated) by Burpee employees as a means to produce random and unpredictable change. Consider an embellished description of a method of zinnia breeding that appeared in the *Saturday Evening Post* in 1951. Zinnias were proving to be a difficult flower to mold into new forms. So Burpee arranged to have some seeds "bombarded with X rays at the University of California laboratories." When this produced no results, "he had the zinnia patch fertilized with radioactive phosphorus." Then, in a final desperate move, "he had the trial gardens sprayed with colchicine" (which, according to Taylor, "ordinarily has the same effect on plant life as spraying a New Year's Eve house party with 100-proof Scotch"), and this produced the hoped-for outcome: the material for several new strains of zinnia. One type was indeed larger, as might be expected for a variety with a doubled chromosome number, but the other changes had been less predictable alterations in the plants' physiology, such as "curly-petaled" and "fluffy, ruffled" types.[11] Clearly, both trial-and-error and tinkering dominated, even in the "flower manufacturing" process. The description suggests that Burpee breeders did not necessarily have a clear vision of what method of improvement was most likely to be effective, and resorted to open-ended trialing of various approaches. And even if they had deployed all the modern techniques for generating mutations in a more systematic manner, producing a flower novelty from any of these was still fundamentally a search for "sports" like those once carried out by Burbank—albeit one sometimes aided by creative experimentation with various tools and techniques for manipulating genes.

Though these may not have provided a streamlined process of engineering, there was nonetheless the perceived advantage of speed that came with inducing mutation rather than waiting for it to occur, and that could not be dismissed out of hand. For Burpee, the need to encourage nature to produce new forms more quickly—in his words, to "shock Mother Nature"—resulted in part from the demands of the commercial marketplace. During the 1930s, the United States provided no intellectual property protection for seeded plants.[12] When a new variety was introduced to the market, its seeds could immediately be planted, harvested, and sold by anyone, regardless of who had developed the variety. This appropriation of newly introduced varieties was not considered unethical; in fact, it was standard industry practice. Therefore a successful business model for a commercial seed company, whose products would for the most part grow to be seed-producing and hence self-replicating plants, entailed developing a new variety and then promoting it heavily to draw a maximum profit before others had time to grow and sell it as well. In

time, the originating company would let the plant drop, in price and in catalog placement, in favor of other, newer varieties, over which it had greater
market control.[13] An alternative to this was to develop first-generation hybrids.
These sometimes made for better plants, but more important, the fact that
they were often sterile provided what one Burpee biographer called "a kind of
grow-your-own patent protection" in which seed had to be obtained from the
company maintaining the two parent lines of plants.[14]

The structure of this competitive market encouraged the relentless pursuit of novel varieties, and it prompted commercial horticulturalists such as
Burpee to embrace any process that might accelerate the appearance of new
traits. A seed company's long-term business success depended on a steady
stream of new plants, and waiting around for nature to produce them was a
poor strategy. In 1940, after only a couple of years of using colchicine, a reporter described how "Mr. Little [the head breeder] at Burpee's will show you
a small moonbeam petunia which after colchicine treatment produced huge
leaves and flowers. A snapdragon, instead of growing blossoms on spikes in
the normal manner, produced flowers irregularly above the leaves all along
the stalk. An inch has been added to the blossoms of zinnias. A large orange
cosmos blossoms three weeks ahead of normal."[15] Such results, or rather the
promise of such results, helps explain the zeal for colchicine demonstrated by
Burpee and others. It was not necessarily that it enabled a deliberate breeding
plan to unfold but rather that it offered many new options, seemingly all at
once. In other words, although the explicit benefit envisioned of colchicine
had been a more predictable and controlled production of plant varieties, even
professional breeders did not always put it to work in this way. What Burpee
Seed needed, more than "genetics engineering" or "non-Mendelian methods
of plant breeding," was simply new plant varieties, and quickly. The route
through which colchicine produced these changes did not matter.

David Burpee also oversaw the application of colchicine in more planned
breeding endeavors. For example, in the early 1930s, he had hit upon the idea
of hybridizing two species of marigold: the tall, yellow African marigold with
the smaller red French marigold. He hoped that the pairing would result in a
tall red marigold flower, something otherwise unknown. If the idea seemed
straightforward in principle, it certainly was not in practice. Burpee breeders
encountered numerous problems along the way, starting with the challenge
of cross-pollinating by hand the composite flowers of the marigolds, crosses
which often did not take. With the help of a geneticist at Bucknell University,
Burpee did achieve a hybrid—the tall red marigold he had hoped for—but it
was sterile and therefore could not be used as the progenitor of a whole new
line of flowers. Although Burpee would have been content to sell a sterile first-

generation hybrid, the process of hand cross-pollination was too arduous and unreliable to be carried out at scale. Burpee tried various routes to producing the African-French hybrid seeds in economical ways, but it proved impossible to get the cost of production below what customers were willing to pay for the seed. The solution to this problem arrived in 1937: colchicine. At some point during these many trials, Burpee had learned that the reason it was so difficult to pair the African and French species was that they differed in chromosome number. By using colchicine to double the chromosome number of the African variety, Burpee breeders were able to create flowers that would readily hybridize, without all the labor and expense, and therefore to create a more economical version of Burpee Seed's Red and Yellow Hybrid Marigolds.[16] In short, Burpee used colchicine in more deliberate ways, too, hoping to create new types according to a preconceived idea.

It bears noting that the Burpee breeders' various techniques for using the chemical were not always accurately conveyed in news reports, especially when David Burpee was the primary source of information. Burpee's promotional strategies are said to have been inspired by the nineteenth-century American showman P. T. Barnum, notorious for his flamboyant displays and clever hoaxes.[17] Although Burpee might have declared in print that colchicine was good for causing a "bust-up"—an event in which many individual plants in a large planting would suddenly display new and strikingly different characteristics—his employees characterized the process more soberly. An account of "tetraploid flowers" written by the "men at Floradale" describes a systematic search to discover in which plants chromosome doubling would produce desirable changes, and trials of various treatment methods (i.e., immersion, spraying, application via a salve) to see which might be the most effective.[18]

It is evident, then, that colchicine could be used to create new varieties in more and less deliberate ways. And there was still another use for the chemical at Burpee Seed: advertising. In David Burpee's mind, the science behind the creation of his various plants was as crucial to their commercial appeal as their size, or color, or vigor.[19] As he explained of his customers in 1933, in the midst of the Great Depression, "Even in hard times they will pay several times the usual price for something very special and unusual."[20] So Burpee took to promoting the very special and unusual methods that led to the production of the tetraploid flowers, or indeed any variety he produced, just as much as he emphasized their advantages in the garden. Two years after putting on the market his Tetra marigold, the "first ever flower created by a chemical," Burpee released his "X-Ray Twins," flowers whose origins could be traced to "the same day in 1933 and . . . the same X-Ray treatment."[21] And he was glad to have one of his breeders describe the creation of the company's "Skyscraper" snapdragons in the late 1940s as a process of "many hybridizing crosses, synchronized with shock treatments by chemicals, [which] sparked a series of mutations."[22] The giant snapdragons that resulted were, according to the 1949 Burpee catalog, "a new 'scientific' creation."[23] Such stunts may have produced new varieties, but Burpee hoped that they would improve sales as well.

It is not surprising that at least some of Burpee's midcentury customers would be drawn to these techno-scientific sales pitches. After all, many American gardeners were experimenters and tinkerers, too, testing not only the latest in plant varieties but also growth stimulants, chemical pesticides, and garden tools. This apparent appetite for the latest science and technology linked growers who simply purchased colchicine-improved flower seeds to those amateurs who tinkered with colchicine themselves. All were invested, albeit to different degrees, in a vision of a technology-driven future for garden flowers as much as for any other product, a future in which tinkering with tools and tinkering with chromosomes—whether done at home or left to professionals like Burpee—would be an important route to improved varieties. Historians have used the idea of "technological enthusiasm" to characterize the apparent affinity of many Americans for novel technologies, especially as their presence in daily life increased steadily from the late nineteenth century onward. Science-and-technology hobbyists are often taken as exemplifying such enthusiasm.[24] The concept seems apt here, too. And it is worth noting that this technological enthusiasm extended both to the products made *for* flower-and-vegetable gardens and the products that came *from* them.

Burpee breeders used colchicine consistently from the 1940s onward, in various ways and with varying degrees of deliberateness, to produce and sell flower novelties. The company was in all likelihood the most visible user of colchicine in the United States as a result of its advertising and sale of tetraploid varieties. Its varied uses of the chemical—ranging from tinkering with genes to undertaking more methodical breeding projects to cultivating an image of technological sophistication—therefore matter. Understanding that colchicine could have all these uses, and not just those prescribed in 1937 (and earlier) by geneticists and breeders, is essential to understanding the extent of interest in, speculation about, and experimentation with the chemical by mid-twentieth-century Americans.

By the early 1950s, the colchicine craze had for the most part blown over. Colchicine methods had produced a few products—most notably the tetraploid flowers sold by Burpee Seed—but new colchicine-derived crops were still in the pipeline. The long-desired fertile hybrid of wheat and rye, for example, was first produced using colchicine methods in the 1950s, and a start was made on its being bred for agricultural uses. It took until the 1960s for one variety of the hybrid grain, called triticale (referring to the genus names for wheat and rye, *Triticum* and *Secale*), to finally enter commercial production. Unfortunately, the earliest varieties of triticale had many traits that rendered them unsuitable to both growers and consumers. Another decade of breeding efforts resulted in improved lines of triticale and interest in the crop resurged in the 1980s. Triticale nonetheless did not become a global economic crop as breeders had once envisioned.[25]

As the triticale story suggests, colchicine did not provide the agricultural shortcut for which many had hoped, though it did add another important technique to the array of methods used by plant breeders. After a new type had been created through chromosome doubling, breeders still had to use traditional methods of selection, crossing, or pure-line breeding to produce plants suitable for agricultural production. Longstanding approaches to plant improvement, including time-consuming trial-and-error practices, prevailed. With colchicine, as with x-rays, the exception to this rule appeared to be in horticultural development. Colchicine-derived flower novelties could be turned out much faster than agricultural crops, and quite a few of these appeared on the market. These included flowers bred by Nebel and Ruttle, who contributed to the production of several tetraploid types that Burpee Seed sold in the 1940s and 1950s.[26] Blakeslee, too, found success with flowers, although this came via a drawn-out procedure. He spent more than forty years developing the black-eyed susan, a common wildflower, into a marketable garden flower. He first selected for size, season after season, and subsequently used colchicine-induced tetraploidy to increase the size of the plant still further. After his death in 1954, the rights to the plant were purchased by Burpee Seed, which eventually bred that stock into a line that David Burpee christened the Gloriosa daisy. These daisies proved to be some of the most popular Burpee flowers of the 1960s.[27]

Of the earliest colchicine experimenters, only Eigsti produced a true commercial success in crops. This success, like that of triticale and the Gloriosa daisy, also highlighted the challenges of colchicine breeding more than its capacity as a "miracle maker." Eigsti's career had led him back and forth between academia, agriculture, and industry for decades, but he became devoted to a specific breeding project in the later years of his life. His goal? To produce a seedless watermelon. This would be a triploid plant, a cross between a normal diploid watermelon and a colchicine-induced tetraploid. The triploid was sterile—thus, it did not set seed—but if bred properly would be otherwise as good as the typical watermelon. Breeding triploid watermelons had begun in

Japan, in 1939, and the first successful types had been produced there in the 1940s. Inspired by the work, Eigsti formed the American Seedless Watermelon Corporation in the mid-1950s, and the Colchicine Research Foundation as well, entities meant to serve as vehicles for raising money for colchicine-based watermelon production. The venture nearly bankrupted Eigsti a number of times, but he persisted, eventually developing (partly in collaboration with the Japanese geneticist Hitoshi Kihara) a method of cultivation and a genetic line that would reliably produce seedless watermelons in the United States.[28] Sales of seedless melons reportedly were just 1 percent of the watermelon market in 1985, but in subsequent decades the seedless types took off.[29] In 2010, about 80 percent of watermelons grown in the United States were seedless, produced by a version of the tetraploid-diploid hybridization process that Eigsti had worked so hard to perfect.[30]

As in the case of x-rays, the amount of attention directed to the potential of colchicine as a revolutionary technology of plant improvement and the slow trickle of enhanced crops and flowers it produced seem to be in contradiction. How are we best to understand the colchicine craze? It seems clear that for many plant breeders, at least initially, colchicine held out the hope for a precision procedure of chromosomal manipulation, one that would allow for specific varieties to be made as imagined: seedless triploids, enlarged tetraploids, and fertile hybrids from previously infertile crosses. In the mid-1930s, the methods of plant breeding seemed not to have kept up with other areas of technological development. Mendelian approaches were over three decades old and relied on trial-and-error procedures. The production of polyploids promised to provide a different route to the improvement of some species—but the methods of generating polyploidy were often unreliable. Colchicine offered the chance to replace these methods. The old system of trialing many combinations in order to find one desirable one, of making do with unreliable tools for chromosome doubling such as exposure to temperature extremes or centrifugation—of endlessly tinkering with various techniques, and with genes and chromosomes as well—would be left behind for the direct and predictable method of "colchiploidy."[31]

If the history of x-ray-induced mutation highlights the quest to industrialize biological innovation by making plant innovation and improvement a faster-paced and more predictable activity, the history of colchicine breeding—at least when viewed from the perspective of these breeders—reveals the persistence of this aim through subsequent decades. And yet the history of colchicine, especially as it played out through efforts to create such crops as triticale and a seedless watermelon, also serves as a reminder of the important and enduring place of trial-and-error methods even as breeders tried to leave these behind. As they quickly discovered, colchicine created new options for those wishing to improve flowers or create impressive crop hybrids, and even to have some ideal in mind at which to aim. But it did not eliminate the years of selection necessary to stabilize desired traits or to otherwise develop colchicine-treated plants into marketable varieties.

This does not mean that the colchicine technique proved a failure, even in the short term. As the case of David Burpee shows, some professional breeders valued colchicine as an effective technology for tinkering with genes as well as one intended to produce specific alterations. Random changes, if they meant novel heritable variations, were a significant resource in Burpee's corner of the horticultural economy. What's more, using colchicine to create new varieties offered him the chance to advertise his company's technological savvy. In this context, then, as novelty generator and corporate advertising strategy, colchicine proved a useful technology, right from the start. And another group similarly buoyed colchicine's early reception and circulation. For some gardeners and amateur experimenters, the application of colchicine to seeds and flowers was an easy way to try their own hand at tinkering, to have fun fiddling with the tools of genetics and in the process to play around with genes and chromosomes as well. This group saw tinkering as something to embrace, not to displace with "better" methods. These individuals may not have had much of an effect in creating useful or profitable new varieties through colchicine, but they formed a crucial node in the network of colchicine experimenters that developed in the middle decades of the twentieth century, influencing discussion, development, and dissemination of this technology. The idea of colchicine as a cutting-edge tool persisted longer among expert amateurs than professionals, likely because its accessibility, ease of use, and immediate effects seemed to continually generate new interest—and, on occasion, new plants.[32] At the end of the twentieth century, it was a technique still discussed in amateur circles as a means to improve plants ranging in kind from magnolias to marijuana.[33]

It was diverse users, and diverse uses, that assured a high profile for colchicine experiments through the 1950s and later. Just as x-rays had promised a way to tinker with the invisible but all-important genes, colchicine presented an opportunity to (as David Burpee would say) "juggle" the chromosomes, to push plants into an evolutionarily important process without having to wait for evolution itself to get around to the task. In this sense, regardless of their sought-after end product, colchicine's diverse users did share an interest in having these genetic changes occur at their whim—that is, in producing evolution "to order."

PART THREE

Atoms for Agriculture:
Evolution in a Large Technological System

On a sunny afternoon in July 1957, Seymour Shapiro of Brookhaven National Laboratory's Department of Biology took a few hours to survey the progress of some ongoing experiments in plant biology. It was a couple of miles from the department to the experimental field, a short drive down a narrow road and past the DANGER signs to the edge of a tall chain-link fence. With a Geiger counter in hand, Shapiro entered the field and began his tour. The initial rows of plants he encountered appeared to be growing vigorously, whether apple trees, holly bushes, corn, oats, tomatoes, blueberries, roses, or even the weeds that sprouted up among these many species. As Shapiro walked toward the center, however, the plants thinned out, grew smaller and more twisted. Some of the trees were already showing autumn colors, months too early. The corn was clearly stunted. Berry bushes produced lumpy, unappetizing fruit. Weeds, too, became scarce. When he reached the center of the field, marked by a nine-foot metal pole, Shapiro considered the patch of lifeless ground encircling it. Everything was proceeding as anticipated.[1]

During most hours of the day, Shapiro could not have entered the field much less walked so near the pole, which usually held aloft a piece of highly radioactive cobalt-60 that scattered radiation across the field (fig. 19). At mid-afternoon, however, his Geiger counter remained fairly still. The cobalt had been lowered into a lead shield belowground, a daily procedure that allowed him and other researchers to access this "gamma field" for a few hours without being subject to lethal doses of radiation. The field was used for a number of in-house projects, as well as the Biology Department's cooperative "radiations mutation" program. Many participants in this latter program hoped that the long exposure to gamma rays emanating from the isotope would generate useful mutations in the fruits, flowers, and trees they had sent to be grown in

FIGURE 19. A researcher works in the gamma field. One can assume that when this picture was taken, the cobalt-60 source had been lowered into the shielded vault located below the central wooden platform. NARA RG 326-G, Box 6, AEC 57-5602.

the field. Having surveyed the field, Shapiro would be able to give an update to some of his cooperating researchers—a group that included arborists at the Brooklyn Botanic Garden, fruit growers from New York State, and flower breeders in Connecticut, among others—on the progress of the plants they had left to his care.

Biologists at Brookhaven had carried out investigations into the use of radiation in plant breeding since 1948, studies that grew up alongside the laboratory's better-known research programs in nuclear physics.[2] Of these various endeavors, the gamma field was perhaps the most striking, but it was only one of a number of radiation-generating tools investigated as potential aids to plant breeding in the 1940s and 1950s. Researchers at the laboratory had access to other technologies including a portable beta-ray irradiator that could be moved around the field and a system for exposing seeds to neutron radiation in the nuclear reactor. Although Brookhaven biologists primarily used these for research in genetics, cytology, and physiology, the technologies were also promoted and used as tools for plant breeding—hardly an expected research focus at a nuclear laboratory.

The twentieth century, especially after World War II, saw the development of what historians of technology call "large technological systems" in domains ranging from energy production and distribution to communication to transportation. Such systems are characterized by a daunting array of interworking parts, which include not only material technological artifacts (in the case of energy production, objects such as coal-fired power plants, transmission wires, home electrical outlets) but also organizations (commercial energy suppliers, equipment manufacturers, government regulatory bodies) and knowledge (physics textbooks, electrician certification programs) and perhaps still other elements, all of which operate in conjunction with one another and are oriented toward the same end goal (the delivery of electrical power). The production of atomic energy in the United States, whether used in power generation or weapons creation, offers another example of such a system. Enrichment facilities, nuclear reactors, regulatory agencies, industrial contractors, national research laboratories, private universities—these many components and still others operated, in whole or part, to advance American nuclear capabilities in the postwar decades.[3]

One feature of such systems, besides their size, is that they tend to foster innovations that perpetuate the system. For example, once significant social and economic investments in an electric power station and grid have been made, an engineer might be reluctant to propose a technical change that would disrupt its operation or entail costly redesign of other components, even if such an innovation would be mechanically more efficient, provide greater safety, or offer some other advantage. The rewards to the disruption or expense would have to be very great. More acceptable innovations would be those that straightforwardly allow for continued production, or extension of the system, even if these were by the engineer's standards suboptimal. Other desirable innovations would be those that enable greater consumption of electricity by end users, thereby creating demand for greater power production.[4]

It is possible to consider this phenomenon on a still larger scale, this time with reference to atomic energy. Once a costly national infrastructure for producing atomic energy and other atomic products is in place, it might prove difficult to dismantle or redesign various components, even when new dangers and objections arise. At the same time, it becomes increasingly important for institutions and individuals within that system to generate technologies that rely on its key product—atomic energy.

Considered from this perspective, it is hardly surprising to encounter the many means of consuming atomic energy that were developed (or redesigned or newly promoted) within the American nuclear system, especially during its early decades. Atomic weapons, nuclear submarines, domestic power stations, radiation therapies, radioisotopes for experimental use—even gamma fields and portable cobalt-irradiators for plant breeders—these were all arti-facts generated within a technological system that also provided grounds for its continued existence. The atomic infrastructure demanded, and produced, atomic innovations. And through the spread of atomic-age breeding technologies, it produced biological innovations as well.

Radiation Revisited

The idea that various forms of radiation might be helpful to breeders had been around for at least two decades by the time the cooperative "radiations mutation" breeding program at Brookhaven National Laboratory was up and running. But because very little in the way of improved plants—or even useful traits—had been produced in various x-ray breeding experiments by the early 1940s, attention to these and other forms of radiation as potential tools of plant breeding had all but disappeared in the United States.[1] It took the scientific and technological developments of World War II to reverse this trend. The massive government-funded research and development program that had produced the atomic bomb in wartime also produced new sites for atomic-energy-related research, not to mention interest in nuclear techniques among scientists in many disciplines. After the war, oversight of the development and use of atomic energy transferred from the military to a civilian Atomic Energy Commission (AEC), whose members soon found themselves responsible for managing these myriad new nuclear activities.[2]

From the outset, the AEC sought to create a positive view of nuclear research among Americans. This could not be done by calling attention to the primary purpose of such research, which was to support national security especially through weapons stockpiling, as this only emphasized its inherent dangers. The commission instead aimed to counter fears of and objections to continued military nuclear development by funding research programs that were not explicitly about weapons and which promised desirable outcomes like cheap energy. These included especially research programs in a range of disciplines situated at the nation's new, government-funded national laboratories, a soon-to-be sprawling institutional network.[3] Crucially, it was not just the physical sciences that benefited from the infusion of resources. Because physics-related research seemed inextricably intertwined with the production of weapons, biological and biomedical research became key focal points for government claims about atomic energy as a social good.[4] In short, the AEC and the politicians who backed it fostered an environment in which a research program linked to agricultural improvement such as that established at the Brookhaven National Laboratory in 1948 could flourish in tandem with the on-site development of a nuclear reactor—not in spite of it, but because of it.

It is impossible to make sense of the reemergence of interest in using radiation-induced mutations in plant breeding, and the creation of new technologies and research programs dedicated to this goal, except as elements within a much larger initiative. Technologies such as the gamma field at Brookhaven were not stand-alone innovations. These were components of a large technological system operating alongside and interacting with many other components, all of which were directed at the same overarching goal of securing and advancing US nuclear capacities. Within this system, the radiations mutation program at Brookhaven and its associated technologies came to play an important role: they were offered as evidence of the American government's good faith effort to develop atomic energy's productive capacities alongside its destructive ones.

During World War II, nuclear science was largely a military operation, carried out at an array of government-funded laboratories and military installations to support above all the production of the first atomic weapons.[5] After the United States dropped the bombs, and peace was declared, politicians, military leaders, scientists, and policymakers struggled to work out the best means to deal with the further development of atomic energy. To what ends should research in nuclear science and technology be directed? And, more important, who should oversee this research? Initial proposals that all nuclear-related research would be controlled by the military and governed by secrecy caused dismay among many Americans, including scientists who had been involved in the development of the bomb through the Manhattan Project. Their objections led to a contentious debate about the most appropriate role for the government, and especially the military, in overseeing this troubling yet promising area of scientific research. The debate was settled, at least nominally, when President Harry Truman signed a bill that transferred on 1 January 1947 the oversight of atomic energy from military to civilian control, that is, to the newly established Atomic Energy Commission.[6]

This commission was charged by Congress with responsibility for developing uses of atomic energy in accordance with perceived national security needs, as well as with an eye to "improving the public welfare, increasing the standard of living, strengthening free competition in private enterprise, and promoting world peace."[7] Soon the AEC was doling out resources to a wide range of research initiatives, in biology and medicine as well as in physics and chemistry, to support work that addressed both military and civilian interests.[8] Among the AEC's beneficiaries were several government-sponsored laboratories that had been established during the war, whose administrators and employees sought continued support as they reconfigured for peacetime—though not necessarily peaceful—research. In early 1947, these included two official "national laboratories," Argonne National Laboratory in Illinois and Clinton National Laboratory (soon called Oak Ridge) in Tennessee, and a constellation of other research sites including a laboratory at Los Alamos, New Mexico, and the Radiation Laboratory at Berkeley, California.[9]

A new laboratory, Brookhaven National Laboratory, soon accompanied these. In March 1947, the *New York Times* described the facility, then under construction, as the future site of investigations into "the fundamental forces that hold the universe together and into the mysteries of the vital processes of life"—in other words, research in nuclear physics and the application of the tools of that field to other disciplines.[10] Although there were to be six different departments at Brookhaven—physics, chemistry, biology, medicine, engineering, and instrumentation and health physics—research in the physical sciences dominated the agenda as the laboratory got started in late 1946 and 1947. Plans were put in place to develop and construct an array of tools for pursuing nuclear physics, including two reactors, a cyclotron larger than any built to date, an electrostatic particle accelerator, and other technologies whose expense and complexity warranted their being located at a government-funded collaborative research center as opposed to an individual institution.[11]

Other areas of research came together more slowly. The Biology Department only began to be organized in July 1947, with the appointment of the chemist and physiologist Leslie Nims as its first chairman.[12] Administrators initially envisioned that the department would pursue investigations in three broad areas: the biological effects of radiation, the investigation of biological pathways using radioisotopes, and the development of general research methods in the field.[13] By 1948, there were a few research programs in place to address particular facets of these broad topics, such as the whole-body effects of radiation in mammals, the chromosomal and genetic effects of radiation in plants, and the use of radioisotopes in entomology, botany, and mammalian physiology.[14]

In 1948, the Biology Department hired W. Ralph Singleton, a geneticist and corn breeder from Connecticut, to occupy one of the department's senior-level research positions. Singleton might have seemed like an unusual hire for a nuclear laboratory. Forty-eight years old in 1948, he had worked at the Connecticut Agricultural Experiment Station for more than two decades. There he had pursued research in maize genetics, studying the causes of hybrid vigor with his colleague Donald Jones and corresponding with Lewis Stadler at the University of Missouri about the effects of x-ray and ultraviolet radiation on maize, among other things. He was better known for his work in practical breeding, however, having developed many varieties of hybrid sweet corn for Connecticut growers during his tenure at the experiment station.[15]

Although unusual at the nuclear laboratory, this strong background in agriculture was precisely the reason that Singleton was an attractive candidate to those in charge of the Brookhaven Biology Department.[16] They hoped that someone familiar with the workings of agricultural experiment stations would be able to coordinate activities between these stations and the department, especially as technologies like radioisotope tracers were made available for distribution beyond the national laboratories.[17] Like other senior hires, Singleton was given leave to devise his own research agenda in the area of plant genetics; however, Nims made clear from the start that the research should involve radiation and, ideally, radiation that relied on the technologies available at Brookhaven.[18] Soon after his arrival, Singleton had up and running a project on the genetic effects of radiation on maize.

There were other plant scientists in the department studying radiation effects besides Singleton. His colleague Arnold Sparrow had recently established a research project that considered a whole range of effects that radiation had on plant growth and development (fig. 20). Sparrow had come to Brookhaven in 1947, shortly before Singleton, having worked in the preceding years with Bernard Nebel at the New York State Agricultural Experiment Station on the use of colchicine in producing polyploidy, and later with Karl Sax at Harvard on the effects of x-rays on chromosomes. At Brookhaven, Sparrow focused on determining the comparative sensitivities of various plants to radiation, identifying chromosomal aberrations (visible in slides created of the cell nuclei) as well as more readily observable mutations and morphological effects resulting from irradiation of varying intensities and duration across many taxa.[19]

Beginning in 1949, Sparrow, Singleton, and others in the department collaborated in the development of a novel tool for plant irradiation research: the gamma field. This comprised a piece of cleared agricultural land with a 16-curie radioisotope of cobalt-60 at the center, which was encased in a stainless steel pipe and could be raised (through the pipe) to a position about ten feet above the ground. The premise behind this construction was that the cobalt-60 would emit constant radiation, primarily gamma rays, which would bombard the specimens planted in the field continuously over the entire grow-

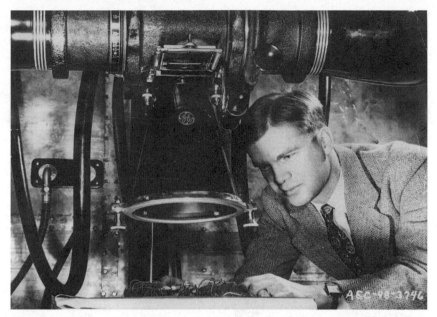

FIGURE 20. Arnold Sparrow of the Brookhaven National Laboratory experiments with x-rays on the wildflower *Trillium* in order to study the effects of radiation on cells. NARA RG 326-G, Box 1, AEC 48-3746.

ing season. Singleton and Sparrow, accompanied first by junior research staff and later by senior colleagues, planted their experimental crops in concentric circles around the source. There they would be exposed to various amounts of radiation, depending on their distance from the center.[20] Because the radiation emitted by the cobalt-60—the "source"—presented a health hazard in addition to providing a novel research technology, the source and pipe had been erected atop a cylindrical lead shield. When a researcher wanted to enter the field, for example, to inspect, remove, or water plants, the source could be lowered into the lead shield by means of a winch located just outside the irradiated area (fig. 21).[21]

In its first season, the gamma field was used to grow maize (for Singleton's research), the common experimental plant *Tradescantia* (for Sparrow's research), tomato plants (on behalf of a researcher at the Connecticut Agricultural Experiment Station interested in the effects of radiation on crown gall disease), and a handful of other species.[22] Singleton expected to see genetic changes among the maize plants nearest the source but nothing among those farther away. The aim of his experiment, as he described it, was to "reveal the amount of constant gamma irradiation necessary to produce a genetic change." Singleton also planned to place seeds and seedlings of corn and barley in specially designed trays that would closely encircle the source to deter-

FIGURE 21. Ralph Singleton, standing at the edge of the gamma field, operates a winch to lower the colbalt-60 source into the lead shielding belowground and watches the Geiger counter to observe the changing levels of radiation. NARA RG 326-G, Box 5, AEC 55-5416.

mine the dose of gamma radiation lethal to a germinating seed.[23] In effect, he was calibrating the gamma field as a research tool. He had virtually no information on what to expect from chronic irradiation—and precious little evidence that such irradiation would be at all interesting from the perspective of induced mutation. Meanwhile, Sparrow sought to determine the physiological and cytological effects of the constant radiation from the cobalt-60 source on his experimental species.[24]

The gamma field exemplified the way in which the expansion of nuclear physics, especially through the increased availability of radioisotopes, shaped the research agenda of the Brookhaven biologists and, by extension, the development of techniques for inducing mutation. Before the gamma field, experimental studies of radiation effects on plants and animals had for practical reasons focused primarily on acute irradiation, such as short exposures to radiation produced by an x-ray machine or a cyclotron.[25] These were, by necessity of the amount of electrical energy required, of relatively short duration. Chronic exposure could have been achieved through the use of radium, a continuous emitter of gamma radiation, except that radium was prohibitively expensive. It had been used in small-scale studies on plant life, especially in the earlier decades of the twentieth century, but it was certainly not suitable for studies that were both large scale and long term.[26]

This changed with the expansion of nuclear physics during and after World War II, in particular with the proliferation of technologies that produced, whether intentionally or as by-products, radioactive elements. For example, in a cyclotron the acceleration of charged particles and their subsequent collision with a target material can be used to generate a radioisotope. Prior to the war, cyclotrons such as those developed by the physicist Ernest Lawrence at the University of California, Berkeley, had been used in this capacity, especially to produce radioisotopes for use in biomedical research and medical therapy; this continued to be an application of the ever-more-powerful cyclotrons built after 1945. Radioisotopes can also be created in a nuclear reactor, through exposure of a target material to a flow of neutrons generated by the fission reaction or by recovering these from the fission by-products of the reactor. In 1946, the US government directed the conversion of the nuclear reactor at Oak Ridge, Tennessee, from its original purpose—the production of plutonium for the atomic bomb—to the mass production of radioisotopes for medicine and research.[27] As a number of historians have charted, the production of radioisotopes after the war, undertaken and heavily subsidized by the US government through the AEC, influenced biological research across the United States and around the world.[28] And just as the production and distribution of radioisotopes is well known to have fostered new areas of medical, biological, and ecological research in the postwar years, so too did the opportunity arise for breeders to take advantage of these as a "new" tool of agriculture.[29] One kind of atomic innovation had spawned myriad others in diverse disciplines.[30]

This process was evident in the gamma field. In 1948, with artificial radioisotopes more readily available, previously impossible large-scale studies of chronic irradiation could now be undertaken. One could think of generating long-term exposures under field conditions as opposed to in laboratory spaces, and over much longer periods of time. Instead of an hour of intense radiation under an x-ray, a plant could be continuously exposed to gamma-ray radiation for the entire growing season, from May to October. Furthermore, because such studies had not previously been done, they promised a potential route to groundbreaking findings in what was by the late 1940s a well-tilled field of inquiry. Singleton and Sparrow, for example, could claim to be pioneering research into the effects of radiation on plants even though such research had been pursued since the turn of the twentieth century.[31]

In 1951, the Brookhaven researchers decided to both relocate the gamma field and increase the intensity of its radioactive source. The new field boasted a 200-curie source of cobalt-60 (as opposed to the original 16-curie source) that had been produced at Oak Ridge National Laboratory. As those who planned the transport and handling of this potentially lethal object reminded Brookhaven staff, "It will be *impossible* for all intents and purposes for any-one to work in the direct beam [of the cobalt-60 source]. Even at ten feet, the *weekly* maximum permissible dose will be acquired in approximately thirty seconds."[32] A few years later, the radiation intensity (which had been increased still further) posed a serious problem when the mechanism for rais-ing and lowering the source failed, leaving the source at its position atop the pipe—effectively preventing any entrance to the field and certainly any effort to approach and repair the snag. The eventual solution decided on was to have the Brookhaven machinist Edward Nicholson, who also happened to be an expert marksman, shoot through the cable from a distance, thereby releasing the source into the lead shielding below. Per orders of the Health Physics Department, Nicholson had to stay at least 100 feet from the pipe, and he was given only one minute to make the attempt. According to a later account, he succeeded on the second shot, liberating the field for use—and avoiding the still worse outcome of striking the cobalt, the dispersal of which would have contaminated the site for years.[33]

In addition to creating new human health threats, the dramatic increase in radioactivity in the gamma field created new opportunities for research. Plants 7 meters from the source would now receive about the same amount of radiation that plants 2 meters from the old 16-curie source had received. This in turn meant a circumference of 44 meters along which plants would receive this dose of radiation, instead of just 12.5 meters—and the expansion of course held for every radiation dose.[34] The scope of experiments expanded in turn. For example, Sparrow took the opportunity to investigate a greater number of species. He now incorporated barley, cotton, lupine, xanthium, snapdragons, gladiolus flowers, tomatoes, and more.[35] The following year, his plantings included nearly fifty different species.[36]

Singleton, too, extended his research. When he was first hired, he had expressed skepticism about using highly energetic radiation in his research program. In 1948, he had maintained to Nims that x-rays only generated chromosomal changes, "translocations and inversions and deletions," and not the more sought-after changes in genes, or "point mutations" as they were known.[37] In this, he aligned with the perspective of Stadler, who continued to insist that the primary effects of x-ray radiation treatment were not gene mutations but gross alterations of the chromosomes that would not be use-ful to breeders.[38] But Singleton's research at Brookhaven evidently led him to reconsider—in fact, to do an about face, and a rather quick one. In his initial negotiations, he had assented to the incorporation of x-rays into his research but only alongside the use of ultraviolet, which he expected would be far more useful.[39] Upon his arrival, this plan was evidently abandoned, as he began almost immediately to explore the use of gamma rays as a means to induce genetic mutations. (Gamma rays are electromagnetic radiation like ultraviolet and x-rays, yet the most energetic and most penetrating of these three.) As a result of his initial studies, which suggested that the rate of mutation in maize increased as a result of exposure to gamma rays, Singleton came to believe not only that gamma rays would induce the desired gene mutations but that they might in fact induce useful gene mutations, and perhaps even be turned into a tool for breeders.

This latter hope inspired a project for inducing a gene for shorter corn. Years earlier, at the Connecticut Agricultural Experiment Station, Singleton had discovered a mutation in sweet corn that produced shorter-than-normal plants. These could be hybridized with traditional types to create plants about six-feet tall instead of the typical fourteen feet.[40] Singleton claimed that the short corn plants were more efficient to cultivate, needing less fertilizer than their larger relatives. The application of this discovery was limited, however, as incorporating the genetic trait into the many different lines of inbred corn then in cultivation would be, to use his words, "laborious and time consuming." If he were able to induce this mutation in many different lines using radiation treatment, however, this would eliminate his having to breed out over many generations the gamut of other unwanted traits introduced in a typical hybridization.[41] In other words, he envisioned a potential use of gamma rays that closely mirrored those imagined for x-rays and colchicine in previous decades: a means of expediting the long, tedious, hit-or-miss process of hybridization and selection.

As he undertook these investigations, Singleton's research began to exemplify yet another aspect of the AEC's increasing involvement in the life sciences, one that went beyond its fostering new research questions, creating new sites for research, and introducing novel research technologies. AEC officials hoped that the commission's commitment to the life sciences would contribute to its stated organizational aims of developing nuclear science for ends such as "improving the public welfare" and "promoting world peace"—or at least convince those watching that it was trying very hard to do so. And this required bringing its involvement in such research to public attention via all available means. From the outset, the AEC and the institutions it supported advertised their life sciences research programs (and other projects, too) through speeches, news reports, conferences, traveling exhibits, and more, in an effort to convince politicians and a broader public of the better world the commission was working to achieve.[42] This was an instance of "technopolitics," in which technologies are supported as means of achieving, or enacting, certain political goals.[43] In this case, the US government, through the efforts of the AEC, supported atomic-related research as part of a larger programmatic effort meant to demonstrate to American citizens, and observers abroad, that atomic energy could be applied to unambiguously positive ends. Researchers, in turn, aimed to show the AEC that they were contributing to this overarching aim.[44]

Singleton's maize mutation research proved a good candidate for such promotional efforts. Coming from a state agricultural experiment station, he was used to touting the benefits to agriculture arising from his research work. This carried over to his new position. In an interview at Brookhaven in the fall of 1949, he proudly showed off the short corn he had begun developing while still in Connecticut. It was, he maintained, an example of how a small genetic change could dramatically improve a crop in a short time—and therefore suggested what the benefits of causing many such mutations in his gamma field might be.[45] A 1951 report on the "atomic cornfield" accepted this premise at face value and described for readers how the gamma field would save plant breeders years of effort. Here Singleton, "as careful with his predictions as he is in approaching his hot cornfield," estimated that the production of desirable characteristics could but cut from ten years to two or three through the use of radiation.[46] Singleton's efforts to tie his maize investigations to agricultural aspirations were further encouraged by his employer, Brookhaven National Laboratory. In January 1952, a laboratory press release announced the production of a "17,000-fold" increase in the rate of mutation in corn, a claim based on Singleton's research. It was presented as an indication that "radiation-induced changes . . . offer the possibility of speeding up the creation of new varieties of valuable food plants," not least by making it unnecessary to search the world for useful genes to incorporate into older varieties.[47]

This Brookhaven statement was released on the same day as the AEC's January 1952 semiannual report to Congress in which the agency highlighted its work to date in the plant sciences. The AEC claimed that the fiscal year ending the previous June had seen $1.3 million allocated to research in the plant sciences, of a total $20.6 million allocated to biology and medicine. The aims of this AEC-funded research were, according to the report, twofold. Some of it was directed toward better understanding radiation effects, work that would aid in developing radiation-protection measures for "life and property." Other research would contribute to the further application of atomic techniques in the plant sciences and foster "projects which may benefit the Nation's people or industries."[48] Research in the area of "radiation and plant genetics" was positioned to contribute to both aims. The description of AEC efforts in this area highlighted the use of radiation to "induce in a short period a variety of mutations that might appear normally only in the course of centuries," an effect that was "an end in itself" for genetics research but which also had "practical aims" in that it could be the source of new varieties. Singleton's research on short corn was, not surprisingly, the key example of work in this area.[49] And it was surely considered a more plausible contributor to the AEC's stated goal of generating research that would "increas[e] the return per acre and per man in agriculture" than many other efforts detailed in the report.[50]

The short corn project does not appear to have amounted to much, but it was only the beginning of Singleton's exploration of the use of induced mutations in breeding. In late 1951, he proposed that some of the gamma field be given over to studies of somatic mutations in fruit trees, evidently thinking this might be a route to the production of potentially useful variations in fruit crops.[51] Soon plans for a cooperative program, involving researchers beyond Brookhaven, took shape around this very idea. This, too, would prove a useful demonstration of the benefits of atomic energy—in more ways than one.

Mutation Politics

Outside collaborators had used the gamma field in its first two seasons, primarily to conduct research projects based on specific agricultural concerns. In 1950, cooperative researchers included a USDA scientist who had tobacco plants irradiated in hopes of producing a mutant plant with, among other things, greater nicotine content. A geneticist at the Georgia Coastal Plain Experiment Station sent pearl millet to Brookhaven for treatment, seeking a line of the plant that would be short in stature, much like Singleton's corn.[1] These and other early collaborations appear to have arisen through personal contacts or individual initiative. Such collaboration was formalized in December 1952, when the Brookhaven Biology Department invited researchers from the USDA and many agricultural experiment stations and agricultural colleges on the East Coast to a conference at which further possibilities for cooperation were explored.[2] The conference quickly led to an official, and far more extensive, cooperative program. Launched in the spring of 1953, the program brought together the nuclear technologies of Brookhaven and the expertise of agriculturalists stationed elsewhere in order to evaluate "the feasibility of producing useful mutations in plants by means of ionizing radiations" via experiments that used both the gamma field and other radiation facilities.[3]

The cooperative induced-mutation breeding program and associated agricultural research at Brookhaven in many ways reflected the origins, organization, and aims of the laboratory as a whole. For one, the research program initially emerged from a need to conduct biological research linked to the laboratory's nuclear facilities—and this was an agenda that studies of the mutagenic effect of radiations addressed directly. In addition, because the laboratory was also meant to be a hub for nuclear research in the northeastern United States, administrators were keen to develop programs and technologies that would engage scientists at other institutions. Inviting outside researchers to have biological materials irradiated at the laboratory facilities fit the bill.

As a means of promoting cooperative peacetime nuclear research centered on unique, large-scale research facilities, the radiations mutation program was clearly a good fit for the still-young Brookhaven. But the specific institutional context that encouraged the development of this cooperative induced-mutation research, and with it interest in what would come to be known simply as "mutation breeding," was itself a product of the growing technological system dedicated to US atomic development. Within that larger system, the use of nuclear technologies in plant breeding satisfied multiple political needs. Not only did it demonstrate the potential nonmilitary benefits of atomic research and development; it also promised to counterbalance more widespread concerns about the harmful effects of radiation on plants, animals, and especially humans—for even as awareness of the dangers of radiation exposure increased in the postwar years, debated among scientists and fretted over in public, plant breeders could rightly claim that not all radiation-induced mutation was bad. After all, that very process might prove key to improving important crops like oats, wheat, or soybeans. Such work offered glimmers of hope for potential benefits of radiation exposure in the midst of an ever-expanding cloud of radiation fears—and further reason for the AEC to embrace this area of research.[4]

The cooperative radiations mutation program initially focused on using the gamma field to produce somatic mutations, preferred because they could easily be propagated asexually, in trees and shrubs. It quickly expanded. Soon Brookhaven was offering seed and pollen irradiation as well, treatments intended to create genetic mutations, and providing access to a greater range of its radiation facilities. Collaborating researchers interested in the effects of chronic irradiation could have plants placed in the gamma field by the Brookhaven staff, where they would be cultivated for one or several seasons and then removed and returned to the cooperating researcher for continued observation. Plants could also be treated in a small greenhouse that had been converted into a cobalt-60 irradiation facility.[5] For those hoping to explore other kinds of radiation besides gamma rays, Brookhaven biologists had access to a number of radiation-generating tools, ranging from x-ray tubes to the nuclear reactor itself, and these, too, were made available to collaborators.

Of these additional radiation sources, the "thermal neutron exposure facility" at the Brookhaven nuclear reactor was the most popular. The facility, also called the "thermal column," was used to study the effects of neutron radiation—that is, radiation composed of free neutrons produced through nuclear fission. It enabled biologists to subject whole plants, seeds, shoots, cuttings, or scions to neutron radiation as it was produced within the nuclear reactor without disrupting the reactor's operation.[6] In the mid-1950s very little was known about the effects of thermal neutron bombardment on plants, and the Brookhaven biologists were eager to discover and chart these effects. They were also more than willing to share access to this resource. Those running the cooperative program used the thermal column to irradiate seeds of some sixty species for more than one hundred researchers across the United States between 1953 and mid-1956—hoping, despite their lack of data, that it might prove useful to breeders.[7]

The cooperating researchers for whom members of the Brookhaven Biology Department performed these "service irradiations" were located for the most part at not-for-profit academic and agricultural institutions. Researchers who used the gamma field in the first years of the program included staff of the Brooklyn Botanic Garden, the Ornamentals Research Laboratory of Cornell University, the Boyce Thompson Institute for Plant Research in New York, the Connecticut and New York Agricultural Experiment Stations, the USDA, and various university academic departments. The plants they had treated ranged widely. The New York State Agricultural Experiment Station presented grapes and apples for irradiation. A Vermont biologist provided sugar maples. The Ornamentals Research Laboratory had ten species in the gamma field, including chrysanthemums, lilacs, and oak trees. Others contributed blueberries, peaches, tobacco, holly, spruce, pine, and more. This pattern held for collaborators using other radiation facilities as well, with university departments, several state agricultural experiment stations, the Rockefeller Institute for Medical Research, the South Carolina Wildlife Resources Department, and the USDA all contributing seeds, scions, or other plant material for irradiation. Species treated included vegetable crops like squash, broccoli, tomato, and lettuce; important agronomic crops such as wheat, rye, oats, barley, and tobacco; forage crops; and even garden flowers.[8]

The success of the Brookhaven staff in attracting collaborators proved to be a boon to administrators of the laboratory as whole. Above all, it indicated that Brookhaven was achieving its goal of providing unique facilities that attracted researchers from across the northeastern states—not only in its physical sciences programs, but also in the life sciences. The 1954 annual report of the laboratory, which emphasized the expansion of collaborative research ("one of the original objectives in establishing Brookhaven National Laboratory"), included the gamma field as one of its four major cooperative facilities, alongside the facilities for which Brookhaven was (and is) far better known—the cosmotron, the cyclotron, and the nuclear reactor.[9] The report proudly boasted that the plant breeding program, "conducted in conjunction with 17 universities and agricultural experiment stations," was now dominating activities in plant physiology: "Almost half the gamma field is now being utilized for this project, and nearly half the time of the thermal column."[10] Although it was not one of the atomic research technologies initially envisioned for this peacetime national laboratory, the gamma field fit right in among Brookhaven's other, more expensive and technologically complex, facilities.

When Singleton had first arrived at Brookhaven, he had been hesitant about the use of radiation in his research program, aligned as he was with the views of Stadler about the nature of changes produced in most radiation treatment. As he transitioned to a new perspective on radiation-induced mutation, evidenced especially in his championing of the cooperative program at Brookhaven, Singleton began to publically dismiss Stadler's longstanding evaluation of the value of induced mutation to breeders.[11] "My views on the usefulness of radiation in plant breeding have undergone a considerable change during the last four or five years," he noted in 1953, referring of course to the time he had been on the Brookhaven staff. "I think that we [maize geneticists] have all been under the influence . . . of Stadler in this country, and Stadler has a definite feeling that all radiation induced mutations are deleterious."[12] By 1953, Singleton felt certain that radiation, including x-rays, gamma rays, and neutron radiation, could be put to effective use by plant breeders.

There were a number of reasons for this new certainty. When describing why attention should be given to radiation-induced mutation, he frequently pointed to the efforts of a group of breeders in Sweden, led by the geneticists and plant breeders Åke Gustafsson and Herman Nilsson-Ehle, who had used radiation, chiefly x-ray and ultraviolet, to produce mutations in barley during the 1930s and 1940s. By the 1940s, their many induced-mutation variants included types with higher yields, altered stiffness, and larger seed size.[13] These served as ready examples of the potential for creating useful genetic variations, particularly in the absence of evidence from US breeders.[14] But Singleton was also influenced in his opinions by the ongoing research at Brookhaven. He was enthusiastic about the high rates of mutation he had found in maize grown in the gamma field. In 1953, he gained further confidence from a breeding experiment conducted by his Brookhaven colleague Calvin Konzak in which exposure to thermal neutrons appeared to have produced an oat variety resistant to rust (a fungal pathogen notoriously damaging to oat crops).[15] Then Singleton himself, in collaboration with a graduate student, found what he understood to be several somatic mutations among carnations grown in the gamma field. The most striking of these was a carnation of the White Sim variety, typically expected to bear white flowers, which bore wholly red flowers instead. He subsequently offered this apparent somatic change as further evidence that cobalt-60 would be an important tool of plant breeders.[16]

As interest in the application of induced mutation in breeding began to pick up at institutions across the country, some geneticists took issue with the growing attention given to the idea.[17] In 1953, Joseph O'Mara of the Iowa Agricultural Experiment Station expressed dismay to his fellow geneticist Ernie Sears at Missouri over the changing focus of the plant breeding program at his institution: "We are in a forest of radiation of oats because a couple of young geniuses—Konzak at Brookhaven and [Kenneth] Frey at Michigan . . . discovered rust resistance in some irradiated oats in some extremely uncritical experiments."[18] O'Mara was equally dismissive of the claims made by the Swedish researchers about the improved qualities they had produced in a range of crops.[19] Both Sears and O'Mara had worked with Stadler at Missouri in the study of genetic mutation and knew well that external factors could affect an experiment so as to suggest higher incidences of mutation than actually occurred. O'Mara clearly shared Stadler's skepticism that any aid to plant breeding could ever emerge from the induced-mutation radiation studies, and there is evidence to suggest that Sears felt similarly.[20] "All that he knows about neutrons 'I could write on one as it went by,'" despaired O'Mara of one new convert to radiation genetics.[21] Such opinions reflected an apparent divide between genetics researchers, some of whom obviously did not consider the surge of induced-mutation research at Brookhaven and elsewhere in a favorable light. They objected that the research was faddish, conducted without adequate knowledge of all the variables involved and with far too rosy a view of the eventual outcomes.[22]

Still, for a time, interest and enthusiasm arising elsewhere advanced the cause of induced-mutation work regardless of the prevailing opinion among geneticists and breeders. This arose in part from the willingness—imperative, even—of the AEC to provide support for biological research to encourage positive assessments of atomic development. The agricultural research at Brookhaven, with its demonstration of an immediate potential application of atomic energy to a national need like agricultural production, was an obvious go-to example for AEC officials touting the benefits of investment in nuclear research to Americans and the world. These favorable conditions were already at work in the early years of studying the biological effects of radiation at Brookhaven, and no doubt influenced Singleton and his colleagues as their research programs developed. But the imperative for work such as theirs further intensified with articulation of an official government agenda for promoting "peaceful uses" of atomic energy. In December 1953, President Dwight D. Eisenhower addressed the General Assembly of the United Nations to advance a set of ideas that would become known as the "Atoms for Peace" program. Eisenhower called for a new direction in atomic development, a shift away from the brinksmanship of weapons creation among world powers and toward a cooperative future in which atomic energy would be used "for the benefit of all mankind." He proposed an international agency to govern atomic energy, which would oversee the distribution of uranium and fissionable material—donated from the atomic leaders, then principally the United States and the Soviet Union—to be used in so-called peaceful pursuits. As Eisenhower envisioned, through the distribution of these materials, "experts would be mobilized to apply atomic energy to the needs of agriculture, medicine, and other peaceful activities."[23] In making this proposal, Eisenhower and his administration sought to advance the military and political interests of the United States as much as international cooperation and scientific and technological development. They especially wanted to divert attention from the government's commitment to producing ever more (and ever-more-advanced) atomic weapons and to counter negative publicity about weapons testing.[24]

The radiations mutation program at Brookhaven easily aided in promoting this agenda, both in the United States and abroad. As part of one outreach effort in the spring of 1954, Singleton was invited to participate in congressional hearings on the uses of atomic energy in agriculture. The hearings, conducted by the Subcommittee on Research and Development of the Joint Committee on Atomic Energy of the US Congress, followed on from a similar set of hearings held the previous year that had highlighted atomic power, another commonly touted public good arising from nuclear science. These and other activities, which circulated in newspapers and other public reports, were to no small extent a part of the ongoing publicity campaign. What better area (other than producing energy more cheaply and curing cancer) in which to claim benefits arising from atomic research, and from government funding of the same, than in the production of higher-quality, lower-cost food? From the accounts given at the 1954 hearings, there appeared to be a vast array of atomic applications that could be incorporated into agricultural research and production in the near future for the benefit of all Americans. Over two days, the committee heard about the use of radioactive tracers in the study of plant biology, nutrient cycles, animal metabolism, and soil fertility; the sterilization of food through nuclear irradiation; the destruction of plant pathogens through irradiation; and the use of radiation in plant improvement.[25]

Singleton, who was the key spokesperson for this last topic, warmed to his task of showing how his genetics research would contribute to the greater social good. He predicted that "the science of radiation genetics will soon become one of the most important events in the history of agriculture." By

FIGURE 22. Ralph Singleton searches for mutations in kernel characteristics induced through exposure to radiation in the gamma field. As the AEC caption (printed on the back of the photo) tells us, kernels in the larger pile were "unchanged by radiation," while kernels in the smaller pile show mutations for "color" and "sweetness." NARA RG 326-G, Box 5, AEC 55-5265.

way of explanation, he discussed Konzak's work using neutron radiation to create rust-resistant oats, a success achieved in one-and-a-half years and at "a very small cost" that "would have taken at least 10 years by conventional plant breeding methods . . . at considerable expense," and described a new breeding effort of his own, one designed to produce a strain of corn resistant to leaf blight through induced mutation.[26] Singleton presented the general research program behind these efforts as one intended to elucidate how most effectively to produce mutation, and of course he mentioned the cooperative program, which was already putting this knowledge to work. In summing up this whole range of activities, Singleton emphasized that plant breeders were "on the verge of a new era" thanks to the increased production of radioisotopes and other forms of atomic radiation and to the research programs that put these to use (fig. 22).[27]

Such assessments would in turn be put to use by AEC officials in their own promotional activities. In one striking later example, the AEC chairman Lewis Strauss described the research at Brookhaven to those gathered in Peebles, Ohio, for the 5th World Plowing Contest, an international plowing competition and exhibition held in September 1957. According to one account, Strauss told the crowd that a new tough and high-yielding strain of oats developed by "nuclear scientists at the Brookhaven Laboratory" would mean "a potential savings of $100,000,000 annually to oat growers." And that was not all. Experiments to create "a bush-like corn" that would have "more ears per stalk" promised similar payoffs in that grain, while an innovation in breeding method—hybridization enabled by radiation treatment—would be a means of producing drought-resistant varieties in many different crops.[28]

Singleton's 1954 testimony, which was covered in the national news, drew attention to another important component of the Brookhaven mutations program and the role it played for the AEC. One round of questions pursued a concern seemingly far removed from agricultural production: Representative Carl Hinshaw, the chairman of the subcommittee, asked Singleton to draw a comparison between the radiation levels he used and those resulting from the fallout of the atomic bombs dropped on Japan in 1945. When Singleton noted that the radiation from the bomb would have been far less, Hinshaw took the opportunity to share with the rest of the room his knowledge of the genetic effects of atomic fallout. "I know there has been some concern that there might be a fallout in Japan of enough to do some damage and I do not believe that that is possible," Hinshaw began. During the interview, Hinshaw noted twice, in rapid succession, that no "noticeable mutations" had been found as a result of the bombings, and that the chance of producing such a mutation was "apparently . . . quite small."[29] Clearly, he did not want Singleton's claim to effective and efficient production of mutations to be read as damning evidence against the testing or use of atomic weapons.

That a presentation of the plant mutations research at Brookhaven would lead to a discussion of the effects of fallout from atomic bombs on human beings points to the degree to which the risks of radiation exposure were both in the public eye and an increasingly urgent political issue by the mid-1950s. Concerns about the potential dangers of atomic radiation first appeared in the late 1940s and received attention in both scientific and political conversation and in popular culture.[30] While scientists debated the genetic consequences of the nuclear age within such forums as the National Academy of Sciences, popular productions such as the film *Them!*, which depicted giant mutant ants resulting from atomic testing, transformed radiation-induced mutation into a pop culture concept.[31]

Much of the early discussion of radiation effects among geneticists centered on the changes that would be seen in humans, such as in the victims of the atomic bombs at Hiroshima and Nagasaki. Beginning in 1947, the AEC—through the work of the Atomic Bomb Casualty Commission—engaged in a long-term study of genetic effects (i.e., mutations) found in the atomic bomb survivors in Japan.[32] The study served political ends as much as intellectual ones, responding to and containing fear and disapproval of the bomb while also seeking to understand the effects on humans of exposure to radiation from a nuclear weapon.[33] Plant biologists also contributed to a general awareness of the potential genetic effects of radiation. Investigations of biological material from plants exposed to US atomic blasts at Bikini Atoll in 1946 turned up "rare" and "unusual" mutants, including a "glowing corn" that shone a cold florescent blue when exposed to ultraviolet light.[34] The plants displayed chromosomal derangements to such an extent that they led the AEC to recommend that those exposed to atomic radiations "refrain from begetting offspring for a period of two or three months following exposure."[35] And the bizarre effects were interpreted in news reports as signs of the possible dangers of the atomic bomb to humans. "Men who are 4 inches high, who glow an eerie blue, who can't have children and look worse than any monster seen in nightmares—This could be the fate of the human race if our inheritance is affected by A-bomb radiations in the same way that corn is affected," speculated one reporter.[36]

Interpretations such as this indicated a growing awareness of the potential dangers of radiation exposure, early stirrings that soon burst forth into a highly visible public debate. On 1 March 1954, just one month before the joint congressional hearings on the uses of atomic energy in agriculture, the explosion of a hydrogen bomb in Bikini Atoll by the US government produced a vast and unanticipated shower of atomic fallout. American servicemen were overexposed, as were residents of nearby atolls, and the crew of a Japanese fishing boat within the range of the falling ash fell intensely ill from radiation sickness.[37] The international incident drew attention to the hazards of atomic testing, and the AEC made new efforts both to understand the nature of radiation hazards and to convince the American public—increasingly concerned for their own health and safety—that they were not being exposed to undue harm.[38] Many geneticists were outraged when AEC commissioner Strauss issued a public statement on 31 March assuring Americans that the radiation produced by atomic testing could never be harmful to humans; the subsequent disagreement between the AEC and geneticists, and among geneticists themselves, over the exact nature of the hazard presented by radiation played out in the national media as well as scientific journals in subsequent months and years.[39]

Singleton's testimony, taking place amid the early stirrings of this controversy, featured at the lead of several news reports of the hearings on agriculture where it offered a decidedly different vision of radiation. As one reporter noted, the subcommittee had taken "time out from the H-bomb hubbub" to learn how atomic energy could be used to "improve food production." He highlighted Singleton's red carnation and the high rates of mutation produced in corn in the gamma field as evidence that scientists had figured out how to make better crops more quickly with the help of atomic energy.[40] An article in *Newsday* carried a similar, if more sensational report: "Scientists . . . are using radioactivity to 'speed up' evolution in plant life and may be able to use the same progress [*sic*] to develop a 'superior type' of animal."[41] The message was obvious: far from solely wreaking havoc on living things, as one might reasonably have interpreted from the events of the previous few months, atomic radiation could in fact improve the quality of human life by contributing to the improvement of other species. This was surely gratifying to some AEC officials.

It is clear that Singleton and his colleagues did not need the unequivocal support of the genetics community to continue their investigations into the uses of radiation in plant breeding. The political demand for work such as theirs was significant, and this demand kept the Brookhaven research program afloat in the mid-1950s in spite of criticisms that arose. Not only did it serve as an example of the potential social good arising from AEC-sponsored research and especially the use of atomic energy; it also presented mutations as a positive outcome of radiation exposure at a moment when attention was increasingly drawn to its dangers. This small piece of the growing national atomic infrastructure—including the attempts made to publicize it—contributed to larger efforts to secure the technological system as a whole, ensuring that it would have the support that it needed to continue to grow and produce in spite of potential opposition.

As a consequence, induced-mutation breeding, both as a suite of technologies and as a community of practitioners, began its own growth and expansion. Investment in nuclear science and technology had created the initial opportunity for a resurgence of interest in breeding programs that relied on radiations, especially those generated by nuclear technologies. In turn, this atomic-age approach to breeding promised to bolster support for investment in nuclear science and technology—that is, for continued investment in the entire technological system. And as that system continued to grow, it created still more opportunities for induced-mutation breeding with nuclear technologies to gain a foothold.

An Atomic-Age Experiment Station

In the 1950s and 1960s, the AEC sponsored investigations across the whole spectrum of US research institutions, from national laboratories to private universities, experiment stations to research hospitals. AEC-funded studies covered a range of disciplines and topics—any subject in which the development and use of nuclear technologies or questions about their effects might come up, and even then not necessarily in very direct or obvious ways. Within this expansive research portfolio, the AEC was amenable to funding research on many agriculture-related subjects, and agricultural researchers proved as receptive to AEC funding as scientists working in other disciplines. Agricultural projects ranged from broad, ongoing research programs to one-off experiments, and covered anything from the possible effects of a nuclear detonation on farm crops and animals to the use of radioisotopes as tracers in physiological or nutritional work. The AEC's biology program supported researchers at the USDA, who used radioisotopes to study plant physiology and soil science, and researchers at state agricultural experiment stations working on topics such as the effects of radiation on plant pathogens and the treatment of plant disease via radiation.[1] Although the AEC's role in agricultural research would never equal that which it held in genetics (a discipline well known to have been amply supported by the AEC), it nonetheless shaped patterns of research in various areas of agricultural science.[2]

In at least one case, it also encouraged the creation of a novel agricultural research institution. From the beginning, the AEC's Advisory Committee for Biology and Medicine had recognized the importance of addressing agricultural concerns and tried to foster research in this area. For example, in 1948, the committee recommended that the biological program at Argonne National Laboratory should emphasize "agricultural aspects," given its location in the

American agricultural heartland.[3] Although nothing significant ever came of this recommendation, the AEC shortly thereafter supported the establishment of an agricultural experiment station at another nuclear facility, Oak Ridge National Laboratory, where scientists focused exclusively on nuclear-related agricultural research. This became the University of Tennessee–Atomic Energy Commission (UT-AEC) Agricultural Research Laboratory, the first-ever atomic agricultural research station in the United States.

The UT-AEC laboratory was, of course, another node in the growing atomic technological system, but that was not the only system of which it was a part. The infrastructure for agricultural production in the United States—which included research stations, commercial producers, farm equipment, and agricultural knowledge—can also be considered a large technological system, in this case one aimed at the mass production of American farm goods.[4] Being embedded in this system meant that the work at the UT-AEC laboratory, which was run as a state agricultural experiment station even though it was sited at the nuclear laboratory, took a different form than the research at Brookhaven—it more resembled a typical state-funded agricultural research program. If looking at the cooperative radiations mutation program at Brookhaven suggests how some areas of agricultural research became embedded in the national system for nuclear research and development, taking a look at the UT-AEC Agricultural Research Laboratory reveals how nuclear technologies became embedded in the established US agricultural research system.

Oak Ridge National Laboratory differed in many respects from Brookhaven. Its origins lay not in the postwar expansion of nuclear science but in the Manhattan Project itself. Oak Ridge, Tennessee, had been the site of what was known during the war as the Clinton Laboratories, the first facilities in the world for large-scale production and separation of uranium and plutonium isotopes.[5] The Clinton Laboratories had grown rapidly as a scientific and industrial operation. The US government had acquired the land in Oak Ridge, remote and undeveloped, in September 1942; the atomic pile, later called a nuclear reactor, was in operation within a little over a year; by 1943, there were more than forty thousand people living in the so-called Secret City.[6] As the war drew to a close, the future of the facilities and the many thousands of workers in Oak Ridge was uncertain, but scientists at the Clinton Laboratories and other sites associated with the Manhattan Project successfully lobbied for the facilities to continue as centers of nuclear research and development.[7] At the Clinton Laboratories, postwar work included investigation of and innovation in reactor design, the production and shipment of radioisotopes for use in laboratory research and medical therapy, and the initiation of new research programs in nuclear science.[8]

Another, slightly more unusual, responsibility fell to the laboratory in its early years. Its staff looked after a number of Hereford cattle, most of which had been exposed to the atomic blast at Alamogordo—the first-ever atomic bomb explosion—in July 1945. The US government had purchased an entire herd of these cattle, many of which displayed obvious effects of their exposure such as open sores where radioactive dust had settled after the test explosion. Part of the herd was then sent to Oak Ridge, in the anticipation that Manhattan Project health scientists could use the animals to observe the effects of direct exposure to atomic radiation and fallout.[9] By 1948, Oak Ridge administrators were looking for a new management regime for the herd. The chief of the Office of Research and Medicine at the laboratory approached the University of Tennessee Agricultural Experiment Station (UTAES) to request the assistance of station staff in managing the cattle. An initial negotiation with the University of Tennessee led to an expanded proposal in which the station would not only partner with the AEC in caring for the herd but also develop an entirely new experiment station outpost at Oak Ridge. The joint program, for which a contract was signed on 11 May 1948, was to involve the application of radioisotopes in agricultural research and the study of radiation effects on agricultural production (fig. 23).[10]

The agreement created a new outpost for agricultural science—the University of Tennessee–Atomic Energy Commission Agricultural Research Laboratory—within an established network of eight state agricultural experiment stations. The laboratory was located on the Oak Ridge Reservation, the site of the national laboratory, but like the other Tennessee experiment stations, it was overseen by the university.[11] This arrangement differed sharply from that seen at Brookhaven: this was not a case of practically oriented agricultural and horticultural researchers being invited to collaborate with the so-called basic research team housed at the nuclear laboratory. At the UT-AEC facility, station researchers developed their own agricultural research projects, sometimes but not always with assistance from Oak Ridge National Laboratory staff.[12]

As a result of this arrangement, the research at the UT-AEC laboratory tended to be carried out and described in much the same manner as other state experiment station research. Station reports emphasized that Tennessee farmers would directly benefit, as they did in the activities of all the agricultural stations. "As the atom chasers uncover new information on life processes other Station scientists apply the information to research in their respective fields. And, as practical results are determined, county agricultural workers

THE TIME is not July 1945 and the setting is not New Mexico, but a scene like this could possibly occur at any point in the United States at any time. If such a disaster does occur, America will be better prepared to cope with its aftermath because of research conducted at the UT-AEC Agricultural Research Laboratory in Oak Ridge, Tennessee.

FIGURE 23. One mission of the UT-AEC Agricultural Research Laboratory was to better prepare Americans for the aftermath of "a scene like this," which "could possibly occur at any point in the United States at any time." From *UT-AEC Agricultural Research Laboratory* (Oak Ridge, TN: UT-AEC, 1966), 2, by permission of AgResearch, University of Tennessee, Knoxville.

of the Agricultural Extension Service pass along improved practices to farm families," described one 1954 report.[13] At Brookhaven, by comparison, annual reports emphasized that the Brookhaven researchers themselves were neither conducting agricultural research nor perfecting seeds and plants for release to the market. They were conducting research in genetics and merely facilitating the application of their findings elsewhere. "The final development of the seed for commercial application is left to the agricultural experimental stations and others," noted one Brookhaven annual report.[14] This is not to say that Brookhaven biologists like Singleton did not aspire to the production of improved crops—they clearly did. Nor does it indicate that the Tennessee researchers did not understand that they were engaged in research that promoted atomic energy. As the 1954 Tennessee station report noted of the experiments at the UT-AEC laboratory, "these tools demonstrate that the atom can be friend rather than foe in our way of life"—an outcome that was perhaps of more interest to the AEC than to Tennessee farmers.[15] The difference was more in the emphasis given to these different goals at the two sites.

Research at the US national laboratories developed under competitive conditions, such that specialization of programs was encouraged and duplication discouraged. The AEC, as the patron of these sites, did not want to waste resources funding two identical research programs.[16] The UT-AEC Agricultural Research Laboratory was not exempt from this need to specialize, and thus to justify its existence within the larger system of AEC-funded research. It was frequently described as filling a particular gap in the expanding portfolio of US atomic programs, being the first to have the capacity to conduct research using large animals.[17] In addition to staff with expertise in animal management and facilities for taking care of herds, the laboratory had specialized apparatus such as a "burro radiation field" where whole-body irradiation of large animals could be carried out.[18] The first studies undertaken at the Agricultural Research Laboratory reflected this specialization, and included studies in farm animals of bomb-radiation effects (i.e., the Alamogordo cattle herd), the metabolism of fission products, the effects of radiation on reproductive function, and radioisotope studies of milk production, alongside other work in topics such as soil chemistry and poultry nutrition.[19]

The effects of radiation on animals remained the central focus of research activities at the Oak Ridge agricultural station for its first five years. During this time, seed irradiation for the purposes of plant improvement took place at the Oak Ridge reactor, but apparently without the involvement of the agricultural station. A few news reports described attempts to induce mutation by "baking" seeds in the "atomic oven" at Oak Ridge. Long trays of seeds (reported to be the same trays used to convey materials for the creation of radioisotopes) were "pushed right into the heart of this pile [reactor], in holes in the graphite and uranium where atoms are splitting on all sides."[20] Among those who sent material to Oak Ridge to be irradiated were the commercial seed company Robert L. Dortch Seed Farms, an Arkansas company interested in the effects on cotton and soybean, and researchers from the Ohio Agricultural Experiment Station and the University of Michigan.[21]

It was not until 1954 that the Agricultural Research Laboratory incorporated plant investigations into its in-house research activities. Thomas Osborne was hired that year as an associate plant breeder in the Botany Department, and he subsequently established a new line of inquiry in plant irradiation. Even before arriving at Oak Ridge, Osborne was well steeped in both radiation research and the challenges of practical breeding. He had completed his graduate study in the Department of Agronomy of the State College of Washington, where an active program in mutation genetics had flourished after the war. The AEC had supported Osborne's thesis research, a comparative cytogenetic study of the effects of x-rays, radioisotopes, and thermal neutrons on various plants. The aim of that project had been to establish a line of wheat that would combine desirable traits of two different types by "radiation-induced translocations"—essentially an exchange of chromosome segments achieved through irradiation.[22]

Osborne continued this pattern of research from his new post at Oak Ridge, focusing on the use of radiation to address particular plant breeding needs. Beginning in 1954, he oversaw research on the improvement of annual forage crops, including lespedezas, crimson clover, and vetches. His approaches for each of these included the same techniques: attempts at "ordinary breeding" through hybridization, exposure to gamma rays to produce mutations, and colchicine treatment to generate polyploidy. Osborne seems to have understood the last two methods as ways to goad the more recalcitrant species into improvement. That fall, he irradiated thirty-eight thousand crimson clover seeds with gamma rays in the hope of finding mutated varieties with traits that would enhance their value as forage plants. As a report detailing the work noted, "The apparent lack of genetic variability in crimson clover, giving little hope of improvement through ordinary breeding, was attacked with colchicine and radiation."[23]

The following year, the station constructed an irradiation facility that was used for these established projects involving plant irradiation and also to support a new cooperative irradiation program.[24] One initial site for radiation treatment of plants had been the station's animal irradiation field (i.e., the burro field). But this could not be used in administering high-intensity gamma rays—a capability needed to treat dormant seeds in particular—so Osborne and his colleagues designed and built the new unit for this use especially. It consisted of two concrete-block buildings sixty-four feet apart. One contained a radioactive cobalt source housed in stainless steel and the other functioned as a control house. The cobalt source was stored at the bottom of a water-filled well in which it was kept for shielding, and a researcher could raise or lower it as needed from the control house by means of a hand crank. Small objects such as seeds were placed in a plastic cylinder that would be completely surrounded by the cobalt source when it was raised (fig. 24), thereby receiving the highest levels of gamma ray exposure; alternatively, experimental materials could be placed on a circular wooden platform that rotated around the outside of the source.[25]

The in-house research program that relied on these irradiation facilities involved studies of the genetic and physiological effects of radiation on plants along with efforts aimed at making induced-mutation breeding practical such as determining the appropriate dose of radiation for various types of seed. Regardless of their specific aims, however, these activities tended to be described using a formula typical for an agricultural station: any research undertaken at the station, no matter how removed from everyday farming needs it seemed,

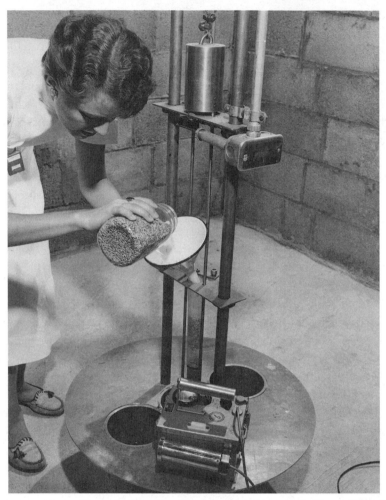

FIGURE 24. A researcher at the University of Tennessee–Atomic Energy Commission Agricultural Research Laboratory prepares seeds for irradiation by a cobalt-60 source. NARA RG 326-G, Box 6, AEC 57-5615.

would eventually inform agricultural practices and therefore benefit farmers. A case in point is Osborne's participation in a soybean investigation, one of the more extensively publicized plant irradiation studies associated with the UT-AEC laboratory. In the 1950s, the soybean cyst nematode was particularly destructive to Tennessee crops, prompting researchers at the state's agricultural experiment stations to respond with investigations into its prevention or eradication. Plant pathologists conducted research to better understand the life cycle of the pest, and breeders sought resistant varieties that might be developed into productive new lines. Following this trend in research activity, Osborne conducted an induced-mutation project in which he exposed soybean varieties to gamma rays from cobalt-60 in hopes of producing useful mutations. Such mutations were to include "resistance to the cyst nematode," in addition to earliness and enhanced seed retention. He declared that "any desirable attributes found will be bred into an improved variety . . . then released to Tennessee farmers."[26] By 1962, Osborne's radiation-based improvement work included, in addition to soybeans, large-scale plantings of irradiated cotton, fescue, and orchard grass.[27]

Osborne's colleagues also participated in the induced-mutation research and similarly directed their attention to projects that would aid Tennessee farmers. For example, Leander Johnson and James Epps of the Knoxville experiment station turned to induced mutation after numerous failed attempts to develop cotton resistant to a particularly harmful fungal wilt. In search of a new approach, they exposed cotton seeds to gamma rays, hoping to produce mutations that conferred disease resistance.[28] In a later cotton breeding project, the researcher Milton Constantine used a portable gamma ray machine, containing a cobalt-60 source, to irradiate cotton bolls after fertilization. He hoped to produce a long sought-after hybrid of American upland cotton and Sea Island cotton.[29] In this case, Constantine did not hope that irradiation of the hybrid cotton would produce a mutation, but rather that it would lead to translocations in which there would be an exchange of segments of chromosomes derived from each of the parent cotton varieties within the cells of the developing seed.[30] If successful, the method would suggest a way to solve similar hybridization problems in other species.[31]

The radiation facilities at the UT-AEC Agricultural Research Laboratory were also used to treat seeds and plants for researchers at institutions across the South.[32] This outreach work had been initiated with the approval of the Advisory Committee for Biology and Medicine, and partly in response to a presentation that Singleton had given to southern agriculturists on the potential benefits of radiation to breeding. It was obvious to the committee that the UT-AEC facilities offered a chance to involve many more southern agricultural researchers in nuclear-related science.[33] The resulting program resembled that at Brookhaven. Collaborators who participated in the program could use the radiation facilities gratis if they agreed to collaborate with on-site researchers by sending in reports on their results; subsequent investigation of radiation effects would be the responsibility of the cooperating researcher. Those who wished to have seeds or other plant material exposed to neutron radiation could arrange to have this done in the nuclear reactor at Oak Ridge, though they had to pay a fee for the service.[34] By 1961, more than fifty researchers had participated in the cooperative program. They represented eighteen different institutions, almost all of which were other agricultural experiment stations in the South. Seeds treated included a wide range of species: major economic crops such as oats, barley, rice, rye, maize, and wheat; forage crops like alfalfa, sweet clover, button clover, orchard grass, and fescue; fruits including grapes, guava, watermelon, and papaya; flowering plants such as forsythia, clematis, daylily, and lupine; and a smattering of vegetables, beans, and other food plants. Some researchers opted for treatments other than seed irradiation. Pollen grains were the most common nonseed contribution, but cooperating researchers also sent roots (sweet potato), budwood (peach), cuttings (grapes and holly), and seedlings (pine).[35]

The UT-AEC staff gathered data—or attempted to—on the outcomes of these irradiations in order to compile a chart of the "relative sensitivities" of the various species and seeds to radiation exposure or, as it was also described, their "radioresistance."[36] This, too, was pitched as a project essential for transforming radiation into an effective and reliable tool for practical breeders. The data produced by cooperators, compiled and analyzed at the station, would be used to inform future breeders about the intensity and duration of radiation to be used for any particular crop in order to achieve the desired balance of genetic change and seed survival. The hope was that this would directly facilitate the uptake of induced-mutation breeding.[37] Unfortunately, this comparative research did not interest cooperators, most of whom dragged their feet on returning the paperwork with their observations. Impatient to gather results, Osborne asked experiment station directors "to nudge your colleagues" into sending in their reports. He emphasized that doing so would benefit their work as well as his own, for "obviously we cannot improve our services if we do not know the outcome of our treatment."[38] Despite this plea, they returned reports for only 30 percent of the material that had been irradiated.[39]

Osborne was disappointed by what he later called "the mortality rate" of those "who dabble in radiation-breeding," a rate he estimated to be about 80 percent. He attributed this high dropout rate to the cooperators' unrealistic expectations of quick and easy results. "Only after they were into it did they realize it was not an automatic, self-adjusting, mysterious, and glamorous system whereby new varieties would somehow spring suddenly into being, needing only to be named and released by the victorious breeder," he lamented. "They expected a sort of 'Instant Varieties' package—just add radiation and stir."[40] Regardless of whether this caricature was fair (and it is difficult to assess in the absence of records from the cooperative program), Osborne and other mutation breeding advocates may have had mostly themselves to blame. Osborne was a vigorous promoter of the use of atomic energy in plant breeding, and in agriculture more generally. His articles in the experiment station's popular journal, *Tennessee Farm and Home Science*, extolled the benefits to farmers of radiation improvement via induced mutation. Though the results would not be immediate, because new types created through induced mutation would have to be crossed back to standard varieties or otherwise developed by breeders for a number of years, he argued that they would no doubt result in valuable plants in time. The use of radiation to produce translocations was similarly valuable in his estimation, a method that "appears promising for such crops as oats; cotton; and hybrids of ryegrass and fescue, sericea and annual lespedezas, and among the true clovers"—in other words, it was promising for many of the species Osborne's Tennessee farm constituents would be most interested in.[41] And Tennessee farmers were not the only audience for this message. According to Osborne, the radioactive sources that had been made available to growers throughout the South, including those provided at the UT-AEC laboratory and others distributed by Oak Ridge National Laboratory, were "potential contributors to agricultural improvements for the benefit of millions of people in several states."[42]

The UT-AEC plant breeding program, which continued into the 1960s, used methods and technologies similar to those innovated and applied at Brookhaven National Laboratory—but its staff directed these toward immediate practical achievements more stridently than did their Brookhaven counterparts. In this the UT-AEC researchers were influenced perhaps by the visibility already achieved by the Brookhaven cooperative program and its claims to some successes by the early 1950s. They were certainly influenced by their particular institutional context, that is, from the establishment of the program as one part of a network of agricultural experiment stations rather than a division within Oak Ridge National Laboratory itself. As such, the mutation breeding program was—like other experiment station research—carried out with an eye to the needs of Tennessee farmers, and with attention given to solving pressing local agricultural problems.

Of course, even if it were not the primary aim of breeders at the Tennessee Agricultural Experiment Station, they nonetheless also participated in the US government's program for promoting the development of—and benefits to—nuclear research, broadly conceived. Poised at the intersection of two technological systems, one dedicated to atomic energy and the other to food production, the UT-AEC researchers found themselves pursuing two distinct aims. On the one hand, their efforts were meant to boost agricultural production. They hoped to produce improved crops for Tennessee agriculturists and to make radiation exposure a more useful tool for breeders. On the other hand, their efforts also supported the development of atomic energy by highlighting for farmers the potential benefits that would accrue from atomic-aided research.

Other agricultural institutions followed a similar path from the late 1950s onward. One telling case is that of the Florida Agricultural Experiment Station, where the AEC supported construction of the station's Cobalt-60 Irradiation Facility. Built in 1958, the five-acre field boasted a 6,400-curie source that would be used to study a whole range of agricultural applications of radiation. As one station report boasted, it was "the largest agricultural irradiation unit in the country."[43] Projects planned for the facility at the time of its construction included the sterilization of fruits and vegetables and the irradiation of meat and animal feed in addition to induced-mutation breeding and associated genetic research. Within the last category, all departments were reported as eager to participate: "The Agronomy department will use radiation to induce mutations in economic crops. The Ornamental Horticulture department will look for radiation-induced changes that produce new horticultural varieties. . . . The Fruit Crops department will seek to obtain radiation-induced mutations for chilling requirement and cold resistance in peaches. The Vegetable Crops department will look for disease resistant mutations in peas and beans."[44] All were hoped-for payoffs that would surely appeal to Florida agriculturists and horticulturists. A similar cobalt irradiation unit was installed at the University of California, Davis, the central site for agricultural research in the University of California system, a few years later. As one report noted, the unit was "a gamma source . . . designed at the Atomic Energy Commission's Brookhaven National Laboratory."[45] Here the focus was to be research on fruit crops specifically, considering anything from postharvest treatments to breeding research. Although it boasted 32,500 curies of cobalt-60, by 1962 it was only possible to claim that it was "one of the largest of its type." Still the "new tool" was "expected to lead to many new areas of radiation research for the benefit of California agriculture and science in general."[46]

The intersection of the atomic and agricultural technological systems did not always require an undertaking as significant as the construction of a gamma-ray irradiation facility. It more often meant simply a researcher or group of researchers who carried out atomic-related investigations amid the usual gamut of agricultural studies and experiments. But it is likely that within these programs, too, researchers responded to the churning of the two distinct technological systems in which they were embedded, using the technologies generated by the atomic research system in pursuit of the goals long embraced by the agricultural research system. This experience did not necessarily stop at the US border, especially after 1955. But before turning to the international spread of gamma fields and other atomic-agriculture installations, it is important to take a look at some of the other domestic spaces usurped into the development and promotion of atomic technology—namely, American gardens.

Atomic Gardens

Amid the spread of nuclear reactors and gamma irradiation units, and ever-growing discussion of atomic-aided agriculture, a few commercial flower-and-vegetable seed companies began testing whether atomic technologies would also be useful in their breeding efforts. One of the earliest adopters of such technologies was David Burpee of Burpee Seed, whose evident fondness for the use of mutagens to produce new varieties (and to advertise his wares) perhaps predisposed him to test out this latest technique. He obtained his own supply of radioactive phosphorus as early as 1950 and, in his words, used it to "fertilize the ground" at Burpee Seed's Floradale experimental farm, hoping to generate variation among the flowers growing there.[1] Other American seed companies similarly took up the use of radioisotopes and nuclear technologies in their breeding programs. The Ferry-Morse seed company reported having "bombarded plant parts with controlled doses of nuclear energy."[2] Ferry-Morse, like Burpee Seed, deployed atomic radiation just as it had used x-rays in previous decades, in hopes of generating a random change that might possibly prove interesting as well as marketable, for example, as a new, "atomic-bred" flower. These seed companies attempted to capitalize on popular interest in atomic energy and its applications, discussing their use of nuclear science as an innovation that placed them at the leading edge of scientific and technological development.[3]

Seed companies that applied atomic radiation to produce flowers and vegetables that could in turn be sold as "atomic-bred" initially overlooked a commercial opportunity that proved to be far more appealing: selling gardeners irradiated seeds or bulbs and inviting them to search among the eventual fruits and flowers for unusual new mutations. "Be one of the first to actually grow unpredictable 'Atomic' Flowers & Vegetables!" was the enticement offered in

FIGURE 25. An advertisement for "Atomic-Energized" seeds and plants emphasizes that they are safe, scientific, and potentially profit-generating entertainment: "Only NATURE knows what will happen, but one thing's sure: *You'll enjoy untold gardening fun!*" From *Chicago Tribune*, 19 April 1962, part 2, 8.

1961 by Dr. Clarence Speas, a rural Tennessee dentist who had recently established himself in the seed business. "Join the world-wide search for new species by safe garden experiments in atomic science!"[4] Whether avid gardener or avid experimenter, one could now take on the role of David Burpee or even Ralph Singleton—Speas's advertisements carried the seal of the Atomic Energy Commission—instead of simply waiting for the experiments of these and other professionals to become products on the market (fig. 25). The notion of growing "atomic-energized" seeds turned out to be a popular one among American gardeners, and Speas's seeds soon sparked imitators, including at Burpee Seed.[5]

Such activities reflected in part the existence of a celebratory atomic culture in the 1950s and early 1960s in which enthusiasm for the potential benefits of nuclear technology persisted even amid growing fears of nuclear disaster and escalating critiques of atomic development.[6] Like the colchicine tinkerers, "atomic gardeners" were keen to explore the uses of new technologies for themselves and to produce their own flower and vegetable innovations in the process.[7] Of course, individuals wanting to experiment with atomic energy in their gardens were far more dependent on professional scientists and the developing infrastructure for nuclear research. "Atomic gardens" were tied up in the same technological system as the Brookhaven gamma field and the UT-AEC plant irradiation facilities. They relied on the same nuclear production facilities and on the government sale of radioisotopes. They all drew energy and meaning from the promotion of an "Atoms for Peace" agenda. In this sense, atomic gardens might been seen as another cog in the larger machine of American atomic production—or, at the very least, an unexpected output.

Clarence Speas was, improbably, the most visible promoter of atomic-aided plant breeding in the postwar years. According to one account, his foray into plant breeding grew out of a conversation with his friend Marshall Brucer, a radiologist at the Oak Ridge Institute of Nuclear Studies. Speas lived not too far from Oak Ridge on a three-hundred-acre property known as Patchwood Farm. As the story was later told, Speas had expressed concern to Brucer about his inability to find grass varieties suitable for his hilly farmland. Brucer, who might well have been aware of the work in radiation-induced mutation being carried out at the nearby UT-AEC Agricultural Research Laboratory, had pointed out in response that there seemed to be a lot of potential for improving plant varieties through induced mutation.[8] Unfortunately, research into the subject was proceeding slowly. The conversation opened the door for Speas's eureka moment: "Why not irradiate a lot of seeds, put them on the market, tell how they should be used, and let private growers aid in mutation research?"[9] Why not invite gardeners into the ever-more-sprawling network of atomic research?

The idea launched Speas into a new career as a seed salesman and brought him into close contact with the American atomic infrastructure. In the ensuing months, he consulted engineers and designed an irradiation facility, which consisted of "a small automated plant" in which "automatic timers propel and withdraw a small car containing seeds, down a concrete lined tunnel and through lead doors to the radio-active target area."[10] He obtained a license from the AEC to purchase and use a radioactive source. With materials purchased from Oak Ridge National Laboratory, he undertook construction of his irradiator, so that in December 1957, when the laboratory delivered to Patchwood Farm a 10-curie source of cobalt-60 packaged in an 800-pound lead cask, his new commercial enterprise could begin. Speas planted the first set of his atomic-energized seeds, irradiated and grown in 1958, to see for himself what effect if any the treatment might have. Then, for the next several years, Speas brought these seeds to a wider market via his new company, Oak Ridge Atom Industries Inc.[11] Interest in the seeds grew slowly, with only a handful of reports making their way into newspapers in 1959 and 1960.[12] In 1961, however, Speas's atomic seeds seemed to be all the rage in home-and-garden circles, inspiring amateur poetry and amateur experimentation, and producing celebration and some condemnation as well.[13]

One sign of this growing interest, and of Speas's increasingly savvy marketing strategies, was the incorporation of the newly invented atomic garden into several annual spring home-and-garden shows held across the country. He arranged for the first such garden to be displayed at the 1961 Cleveland Home and Flower Show, an annual event advertising the latest trends in home-and-garden design (fig. 26).[14] Plans for the one-thousand-square-foot atomic garden display began months in advance. The hope was that it would include mutant plants raised by Oak Ridge Atom Industries and professional growers, as well as plants raised from irradiated seed by local Ohio gardeners.[15] The result would be a display that conveyed both the potential for improvement in flowers, fruits, and vegetables, and the openness of the enterprise of atomic gardening to anyone who wished to attempt it. It would also highlight the as-yet experimental aspects of growing plants from irradiated seed, inviting viewers to compare for themselves the differences between atomic-energized and other plants.

The local newspaper, the *Cleveland Plain Dealer*, collaborated in the effort, offering free irradiated seeds and a pamphlet entitled "Workbook for a Garden Experiment in Atomic Science" to readers who mailed in a coupon declaring their desire to become a "*Plain Dealer* Atomic Gardener" in November 1960.

FIGURE 26. Flowers bloom and vegetables ripen in the country's "First Atomic Garden," a feature of the 1961 Cleveland Home and Flower Show. Photo by Frank Scherschel, The LIFE Picture Collection, Getty Images. Used by permission.

The paper appealed to a wide range of potential motivations for participation, from the opportunity to have a flower on display at the Home and Flower Show, to the chance to aid in the development of scientific knowledge through experimentation, to an economic, or possibly philanthropic, interest in discovering a particularly valuable mutation.[16] Regardless of which outcome most appealed, many gardeners sought out the irradiated seeds. Within a week, twenty-five thousand requests had arrived at the public service bureau of the *Plain Dealer*, a demand that far outstripped the five thousand packets it had to share.[17] In the weeks leading up to the garden show, the *Plain Dealer* reported on the progress of its various atomic gardeners.[18]

By the time the show rolled around in March 1961, Speas had quite a lineup of products. Would-be atomic gardeners could choose from many varieties of irradiated flower and vegetable seed, anything from sweet pea and snapdragon to tomato and onion, or they could opt for irradiated rose bushes or gladiolus bulbs, or simply substitute "atomic-energized potting soil" for the usual stuff.[19] Speas reportedly was on hand at the Home and Flower Show every day from open to close. There, in a booth adjacent to the display atomic garden with its rows of irradiated roses, chrysanthemums, tomatoes, begonias, and other popular garden flowers, he responded to visitors' inquiries about atomic gardening. He no doubt encouraged many of them to try for themselves with the seeds he had for sale on-site.[20]

Other atomic garden displays appeared at home-and-garden shows across the country in 1961 and 1962. The 1961 Los Angeles Home Show boasted an atomic garden, which, like that in Cleveland, included display plants grown from irradiated seed.[21] Even much smaller shows, such as those in Cedar Rapids, Iowa, and Hutchinson, Kansas, featured booths dedicated to explaining and selling irradiated seeds.[22] Some shows aimed for displays that overwhelmed. At a 1962 Maryland show, the "neutron" (probably a model of the atom) was incorporated into a garden setting in the "Atomic-Age Garden." The garden ran from the main entrance to the middle of the auditorium, with a "cluster of pink, blue, and beige neutrons" suspended at the center, while at ground level, dogwood trees, rose azaleas, and purple hyacinths stretched skyward.[23] In other cases, the so-called atom-energized seeds appeared as just one innovation among many that would improve gardens in the season ahead, worthy of mention but not headline news.[24]

As a result of trade shows, news reports, and especially Speas's advertising, the idea of atomic gardening did capture the attention of American gardeners, at least for a couple of seasons. Sensing that they might be losing a potential market, other seed producers signed on to the notion. Burpee advertised in 1962 irradiated marigold seeds and encouraged growers to search among the resulting flowers for an elusive all-white marigold.[25] Breck's Seed, another prominent and similarly long-lived American seed company, offered its own six varieties of "Atomic Seeds!"[26] The radiation trend caused one reporter to speculate about its continued appeal, "Hot [irradiated] bulbs and plants are likely to boom garden sales this year and for many years to come."[27] This reporter turned out to be mistaken, for popular interest in atomic gardening as a backyard hobby had already reached its peak and would drop sharply in the years ahead.

Before the fad receded, however, quite a few people took the plunge into atomic gardening. From the start, the assumed appeal to growers of using irradiated seed was less the promise of prettier or easier-to-manage garden plants and more a shared curiosity about the effects of radiation and the aspiration to produce something novel—circumstances not unlike those that surrounded their use of colchicine. This was evident even in the packaging. Oak Ridge Atom Industries sold its seeds either as individual seed packets, a single packet for one dollar with a second just fifty cents, or as part of an "experimental control kit." The latter cost about four dollars and contained four packages of radiation-treated seeds and four of untreated seeds for comparison, as well as an instruction booklet and charts for keeping an eye on differences.[28] The experimental kit no doubt heightened a gardener's sense of participating in atomic research. But even those who couldn't care less about the creation of scientific knowledge might be tempted to participate in the search for mutant plants. In interviews, Speas described how rose breeders would pay up to $50,000 to the individual who produced a blue-colored flower—surely an enticement to purchase some of his irradiated products.[29] Oak Ridge Atom Industries further offered to purchase from home growers any plants "deem[ed] to have commercial value." And if finding a plant with obvious commercial value seemed like too high a bar for the amateur grower, the company had still another promotion, "The New Discovery Project," in which it promised cash prizes up to $1,000 for "the most unusual plant growth reported." All one had to do was send a photograph of the plant monstrosity and a written description.[30]

Those who reported on the atomic gardening trend tended to focus on the same ideas of amateur participation in science and the potential for reward that Speas and his firm emphasized in sales pitches. Encouraged to see each packet of irradiated seeds as unpredictable and untested, growers could imagine themselves at the forefront of this important scientific and agricultural endeavor (fig. 27). "Everything is still in the experimental stage and no one knows what to expect," emphasized one news reporter, who saw both the "thrill" of experimentation and the "fun" of producing something wonderful;

FIGURE 27. A Tennessee woman considers purchasing Oak Ridge Atom Industries' "Atom-Blasted Seeds." Photo by Grey Villet, The LIFE Picture Collection, Getty Images. Used by permission.

she even reminded her readers to report unusual growth to an expert gardener so that the mutants could be appropriately evaluated. Reports such as this took for granted the claim of Oak Ridge Atom Industries that an amateur might actually produce an important new plant variety.[31]

In this celebration of amateur participation, atomic gardening bore similarities to earlier interest in colchicine and the potential wonders the chemical might produce even when applied by those with no training. The syndicated garden columnist Edna King Mandeville told readers to expect "the creation of advanced types" through the application of atomic radiation to garden plants—but not to feel that they had to wait for scientists to produce these: "There is no real reason why a persistent amateur like you or me cannot share in the discoveries in our own back yard."[32] She followed up with directions on how to appropriately carry out a controlled experiment using the seeds, so as to be certain that any changes observed could be traced back to the treatment.

On the question of what kinds of plants might emerge from the irradiated seed, there was little agreement. Many accounts saw irradiated seed as the starting point of greatly improved types. George Taloumis of the *Boston Globe* shared a vision of future atom-aided gardens in which "not only may flowers be larger, more brilliant and pest-free but vegetables may be tastier, produce more and develop new shapes."[33] Others raised fairly modest expectations: bigger and sturdier plants, and possibly more colorful flowers.[34] Still others indicated more dramatic outcomes. One California newspaper described the production through irradiation of "giant, disease resistant commodities on a commercial scale," such as a sixteen-pound squash and a ten-foot tomato plant producing a reported two hundred tomatoes.[35] Obviously, some reporters felt inclined to celebrate the potential for extreme alterations, just as those who reported on the potential products of x-ray and colchicine breeding had in previous decades.

Although there were some commonalities, the terrain of popular conversation had shifted from those earlier times. In the atomic age, many reporters stressed the potential for oddities, the unusual types, the malformed and unappetizing fruits to be found growing in the atomic garden. "Watch for 'strange growths,'" advised Oak Ridge Atom Industries, which listed dwarfism, gigantism, strange colors and shapes, and stunted forms among the likely effects.[36] To a few observers, this rendered the whole idea a little silly. The garden columnist of the *Chicago Tribune* shrugged off atomic gardening with this straightforward explanation: "The so-called 'atomic' seeds are much more likely to turn out scraggly, malformed, or otherwise undesirable plants than they are new 'wonder' varieties. If you don't object to this, go ahead and experiment with them."[37] Other reporters encouraged readers to think of these scraggly types as part of the adventure or as "conversation pieces."[38] For growers who shared this perspective, creating a noticeable change in the form of a plant was enough to repay their investment in irradiated seeds. The atomic garden was a context in which a deformed or injured organism could nonetheless be a sign of success.

Still other observers viewed the potential for mutant plants in a distinctly less favorable light. An editorial published in the *Milwaukee Journal* expressed concern about how far exactly the radiation experiments would go. Speas's claims to new tomato varieties and supertall petunias were all well and good, but the author wondered whether "maybe experiments should stop right there." As the editorial speculated, "Next they'll create a man-eating rutabaga or sweet corn with teeth that bite back."[39] Even if such unwanted crop varieties were unlikely to emerge from radiation treatment, the seeds remained linked to more threatening by-products of the atomic age, such as the bomb and radioactive fallout, as much as they were linked to its wonders. Like more than one person intrigued by the idea of atomic gardening, a *Washington Post* columnist wondered whether there was any lingering radioactivity from the radiation treatment. He encouraged his wife to go ahead with the experiment regardless, but was left with what he called a "nagging question." "If a bit of exposure to atomic radiation can cause petunia seeds to mature into something grotesque, what will radiation do to humans?" he wondered.[40] One small-town paper offered its own sharp assessment of the atomic-irradiated seeds, quipping, "We notice that one of the seed catalogs advertises 'atomic seeds,' which probably means they blow up rather than grow up."[41] And visitors to a Maryland home-and-garden show in 1961 could hardly have avoided the conflicting views of atomic energy embodied by the inclusion, in the same event, of a display of an "atomic energy flower garden" and "a fallout shelter erected in cooperation with the Maryland Civil Defense Agency."[42]

These comments and reports suggest that the anxieties of the atomic age generated ambivalence about this new garden technology among some observers and disgust in others. They offered counterpoints to the more celebratory attitudes expressed by atomic-garden enthusiasts. The wide range of responses to atomic-energized seeds reflected, at least in part, a larger uncertainty about the benefits and costs of atomic technology in American life. Nuclear development could mean cheaper energy production or improved medical technologies; it might equally well result in global destruction through nuclear warfare or increases in rates of cancer and other human diseases through exposure to radiation and fallout. In the case of irradiated seed, nuclear development might possibly mean bigger and better crops and flowers, but it could also entail deformed, fruitless, worthless plants.

What did those who attempted to cultivate plants from irradiated seeds think? A poll of the *Plain Dealer*'s atomic gardeners in late 1961 suggests that some but certainly not all were satisfied with the experience. Many reported pleasing, if fairly ordinary, flowers. A few thought that their atomic seeds were decidedly better than untreated ones, producing healthier and more richly colored flowers in a number of varieties, taller snapdragons, larger and more prolific tomatoes, and "exceptionally vigorous" pepper plants. "The atomic lettuce was out of this world," wrote one reader, mixing his technological metaphors. Apparent mutations seemed to be few. R. J. Reed of Chagrin Falls reported a zinnia blossom that was partially albino. Two elementary school students observed a slew of changes they felt to be important, such as mutated pepper plants including a very dwarfed variety, taller than normal corn, variations in marigold leaves, and an aster that was half albino. A number of growers had decidedly negative assessments. The horticulture teachers at one Cleveland school garden were unimpressed, noting that the outcome of radiation treatment hardly justified the additional cost. Luther Karrer of the Harvey Rice School summed it up best, "Nothing indicated any mutations or anything better than non-energized seed." The apparent failure of the radiation treatment to produce desirable effects—or any effects at all—did not dissuade every grower from believing in the overall value of the experience, however. Mary Sigafoo of Stow, Ohio, submitted the seemingly negative report that her atomic-energized zinnias were stunted and sickly in comparison with the healthy control plants. Still, atomic gardening appealed to her. "I would be most happy to continue other experiments," she concluded.[43]

The lackluster results produced by many atomic gardeners in their first season, especially in comparison with the bold claims made in Oak Ridge Atom Industries' advertising and elsewhere, no doubt contributed to the rapid decline in interest in irradiated flower and vegetable seeds among consumers after 1962.[44] The firm tried other forms of irradiation in order to turn a profit, such as atomic-energized lawn seed whose vigor would prove itself in "hard-to-grow-grass areas" (one wonders if Speas was still concerned about his lawn troubles) and atomic-energized golf balls, "a unique peacetime product utilizing the powers of atomic energy" for "livelier performance" on the fairways.[45] In 1963, the company patented a process for irradiating grape vines to induce chromosome duplication, a treatment that reportedly resulted in "larger fruit, thicker leaves, greater vigor, increased yields and other characteristics," but there is no evidence to suggest this method was taken up by growers.[46]

In most respects, the gamma field at Brookhaven and equivalent facilities at Oak Ridge and elsewhere were vastly different enterprises from atomic gardens, their home and school counterparts. The gamma field was a sign of government commitment to the development of peacetime uses of atomic energy and a means by which the AEC, with the assistance of a group of biological researchers, could advertise this commitment to Americans and to the world. The atomic garden, on the other hand, was a commercial gimmick, one that drew on shared (mis)understandings of genetics and evolution and a certain fascination with atomic science to generate profits for a small Tennessee firm.

But there was more to link the two than simply their rubbing elbows beneath the broad umbrella of atomic-era American culture, in terms of both material operations and cultural appeal. In a way, they were quite securely intertwined. To the 1960s consumer, the most obvious connection between Oak Ridge Atom Industries and AEC programming was the designation "Oak Ridge" and the seal of the Atomic Energy Commission that appeared in some of the commercial firm's seed advertisements. Oak Ridge Atom Industries was eager to be perceived as part of the official US nuclear science establishment. Of course, the company was for the most part just one of the AEC's customers, having reportedly purchased the license and materials with which to irradiate various plants and seeds. Yet other evidence points to transfers of people and materials between Speas's commercial program and AEC-supported induced-mutation breeding efforts, which in turn suggests that Oak Ridge Atom Industries was more than simply an AEC customer. Most telling, there was a transfer of expertise, evident in the overlap between Speas's staff and Oak Ridge personnel. Thomas Osborne, head of the cooperative seed irradiation project at the UT-AEC Agricultural Research Laboratory, was described in one report as a "consultant" to the effort and authored some of its experimental guides, and James Bacon, who had formerly been an assistant at the UT-AEC laboratory, was "Manager of Radiation Plant" at Oak Ridge Atom Industries.[47]

The transfer of end products, in the form of new crop plants, from AEC-sponsored breeding activities to Oak Ridge Atom Industries suggests further interdependence. The atomic-gardening promoters took every opportunity to describe the irradiation process as one that had produced unqualified successes, referring to an increasingly canonical list of crop improvements thought to be attributable to the application of radiation.[48] At the start, the successes could not include their own varieties, given that only a couple of generations of Speas's irradiated plants had been cultivated. So Oak Ridge Atom Industries instead laid claim to a prior innovation, advertising as its own creation an improved peanut that had been bred by Walton Gregory at North Carolina State University. Gregory, supported by AEC funds, had sent hundreds of pounds of peanuts to Oak Ridge National Laboratory to be irradiated with x-rays beginning in 1949. From these a variety known as NC-4X was eventually produced, which was described as having some advantages in hardiness and other characteristics.[49] By 1960, through pathways that remain unclear, Gregory's NC-4X peanut variety came to be marketed by Oak Ridge Atom Industries as "one of the new wonders of the plant world developed through atomic energy!" Touted as a lovely houseplant, the peanut's advantages in a domestic setting seemed to be limited to its inborn energy—"grows with uncanny speed, from day to day, vibrating, shaking, germinates in about 3 days."[50] Subsequently referred to as the "atomic peanut," despite the fact that it had been bred from x-ray irradiated stock, it was a vehicle for advertising both the success of the AEC-sponsored mutation programs—it was frequently touted by Singleton and others as evidence of a successful induced-mutation variety—and Speas's sensational commercial products.

Just as people and plants bridged the AEC-sponsored efforts and the commercial operation, so too did ideas developed in the course of Atoms for Peace programming make their way into conversations about atomic gardening. Planting irradiated flower seeds seemed to bring the Atoms for Peace mission straight into American homes, something other peacetime nuclear initiatives, such as the application of nuclear science to energy production, had yet to achieve. "In their way, these gardeners will be embarking on an era of atomic gardening, thus helping to put the atom to work in a way that will add to the enjoyment of life," mused one journalist.[51] In fact, it not only brought this peacetime atomic agenda into American homes, but demanded that ordinary citizens take on the role of chief scientist and experimenter, transforming wartime technologies into tools that would benefit the world. "Everyone has an opportunity to play an important part in pioneering for better farming and gardening by reserving a couple of small 'garden laboratory' plots on the home grounds," wrote one home-and-garden reporter.[52] Oak Ridge Atom Industries and others who sold irradiated seed pushed this association still further. "Share in this peaceful development of the atom," encouraged one advertisement sponsored by a Chicago grocery store, calling experiments with irradiated seeds "probably the most important application of peaceful uses of atomic energy in which you personally participate."[53]

In one important way, the aims of Oak Ridge Atom Industries or Burpee Seed were decidedly different from those of their counterparts who explored the use of radiation in breeding at the national laboratories and in other government-sponsored programs. The latter by the 1960s were working to establish a global network of researchers and to develop an effective set of technologies for inducing mutation. They sought to create a recognized professional niche in which they would continue to receive support for their work. Meanwhile, Oak Ridge Atom Industries and other seed sellers broke down the researcher-layperson boundary, suggesting to ordinary home gardeners that they were perfectly competent to identify and profit from mutations found among radiation-treated seeds.

Some members of the professional community of geneticists and plant breeders felt threatened by activities that seemed to portray their own work as simple enough for anyone to succeed at. They asserted that atomic-energized seeds were a commercial fraud and that professional training and dedicated multiyear breeding programs were essential to the successful application of nuclear technologies in plant breeding. Singleton thought selling irradiated seed to the public was a "reprehensible practice," and even went so far as to write a piece for the *New York Times* reminding would-be atomic gardeners that all the "new and improved types" created through irradiation "were produced by competent plant breeders who were trained for the job."[54] In other words, atomic-energized seeds were not a legitimate product of the national atomic research and development program.

The same technological system that had generated interest in and resources for using nuclear technologies in plant breeding at national laboratories (and subsequently enlisted these breeding activities in its further development and growth) had produced, by all appearances as an unanticipated by-product, an interest in atomic gardening. This novel activity in turn enlisted home gardeners, for a time, in the work of that system—imagining themselves as contributors to atomic science and technology, while surely contributing to a larger fascination with and celebration of an atomic-enhanced future. It also infuriated some scientists. But Singleton and other researchers who bristled at the association of their research with the commercial gimmickry of companies selling atomic-irradiated seeds for the home garden need not have worried too much. It proved only a fad, and it blew over quickly. By comparison, "mutation breeding" as a professional research activity, including using atomic-derived technologies, persists into the present.

The Peaceful Atom in Global Agriculture

The 1950s saw the development of a new international atomic infrastructure, one that would eventually be managed, at least in principle, through an International Atomic Energy Agency (the IAEA, created in 1957). In this international system, too, American atomic technologies were explicitly the object of technopolitics. They were supported as means of achieving certain political goals, including in the arena of global development. The earliest US aid programs for building nuclear capacities (especially nuclear reactors) had been aimed at European countries, efforts to expand technical skills and create economic and political ties among friendly nations.[1] But the US government saw that promising the same to newly independent nations around the world was also to its political advantage. It was a bargaining chip to win the allegiance of these countries—perceived as strategically essential in the new Cold War order—and a way to establish American industries abroad. It would also enroll these nations in the still-nascent system of control and surveillance that accompanied possession of weapons-grade fissionable materials.[2] This relationship was push-pull. Countries also sought to develop their nuclear capacity, including through American aid, seeing the deployment of atomic energy to diverse ends as an important demonstration of their ability to stand as independent nations among the global powers.[3]

Nuclear technologies for plant breeding were swiftly incorporated into this growing global nuclear system. Irradiation equipment like cobalt-60 sources were easily exportable, and with fewer attendant political risks than nuclear reactors. Much like the "clean" energy production of a reactor, induced-mutation methods promised modernized, high-tech production—in this case, of new plant varieties. These could be envisioned as well suited to countries whose agricultural systems were only just modernizing along the lines of American industrial agriculture.[4] They were symbols of American influence (and ideally American largesse, too), especially when the AEC donated or made available for purchase the radioactive sources essential to new atomic-age breeding programs. And, from the perspective of acquiring countries, induced-mutation breeding technologies were a sign of having the scientific and technological know-how to manage this modern approach to plant innovation.

As I argue here, if the reemergence of interest in using radiation in plant breeding in the United States depended on the growth of a large technological system dedicated to the development of atomic energy, then its continuance—and indeed the very existence today of "mutation breeding" as a well-defined field—depended on the global extension of that technological system. The AEC and later the IAEA cultivated interest in mutation breeding, especially that which relied on nuclear technologies, through international conferences and also international aid. These activities enabled the spread—and to some extent the survival—of nuclear-derived plant breeding technologies, especially as part of a development agenda, even as breeders in the United States became increasingly disaffected.[5]

In the months following President Dwight Eisenhower's "Atoms for Peace" speech before the United Nations in 1953, policymakers in the United States and around the world worked out what exactly Eisenhower's declarations would entail, from the structure of the new international agency to the scope of its activities to whom would be invited to participate. As part of these deliberations, plans took shape for an international scientific conference that would highlight the potential peaceful uses of atomic energy.[6] From an initial proposal of a fairly small conference hosted by an American scientific organization such as the National Academy of Sciences, plans ballooned. By late 1954, the conference was being organized under the sponsorship of the UN, to be held in Geneva in the spring of 1955. This International Conference on the Peaceful Uses of Atomic Energy was to involve the presentation of technical papers by scientists and engineers from many nations, on subjects ranging from power generation to biological tracer studies with isotopes. It would serve as a forum for the global exchange of information on nuclear-related research.[7]

At the Geneva conference, agriculture took its place as one of the many fields to benefit from nuclear science, alongside power generation, medicine, biological and chemical research, and industrial production. Ronald Silow, a British plant geneticist and a delegate of the UN Food and Agriculture Organization (FAO), noted several areas in which agriculture could benefit from nuclear research. Cheap energy production through atomic power would reduce the energy-related expenses of farmers. It would also enable the development of new areas for food production by reducing the cost of transportation and infrastructure needed to extend irrigation and cultivate more remote regions. Radiation and radioisotopes would make still other contributions: radiation by offering opportunities to improve plants, sterilize foods, and provide pest control, and radioisotopes by aiding scientists in their quest to understand biological processes. In glowing rhetoric that matched the grand aspirations of the conference as a whole, Silow estimated that atomic energy's contributions to agriculture could be equal to, if not quite as spectacular as, its contributions to industrial growth.[8]

Many of the agriculture-related presentations at the conference focused on research programs that would aid food production indirectly and only after prolonged research, such as through enhanced understanding of biological and chemical processes. By comparison, induced mutation emerged (alongside food irradiation for preservation) as a promising route to immediate tangible benefits of the peaceful atom in agriculture. Presentations and papers on this subject—a number of which were contributed by American researchers, most with Brookhaven connections—emphasized that radiation could be a valuable tool for producing variations, providing breeders with new genetic combinations that nature had not.[9] Although most acknowledged they did not know in advance what their radiation treatments would produce, they aspired to a future in which mutations would be precisely controlled. These aspirations were best summed up by the Swedish plant breeder Åke Gustafsson. He described how scientists had already determined that "beneficial mutations" could be induced in all crop plants. The orientation of future research was obvious. "In the next few years," he argued, "we have to elaborate methods for the induction of *specific mutations*, changes in just those *loci* where mutations cause an increase of yield or improve upon other characters of significance."[10]

But what exactly were the benefits that this form of genetic manipulation would confer? A number of participants in the Geneva conference envisioned that continued research into radiation-induced mutation would allow them to perfect their technique such that breeders would be better able to keep up with the changing demands of agricultural production. Gustafsson made this clear, for example, characterizing induced mutation as a way to "modernize" crops that were "old-fashioned in their morphology and anatomy."[11] But the need to redesign plants to accommodate increased agricultural mechanization was only one of a number of challenges that those working on induced mutation in the later 1950s identified as imperatives for their continued research. Breeders also needed to keep up with the constant appearance of new agricultural pests and diseases, problems that seemed to be exacerbated by the methods of modern industrial agriculture. As farmers applied new chemical insecticides, insect populations developed greater resistance to these as a result of selective pressure. Ditto for herbicides and weeds.[12] What breeders needed, and the induced-mutation researchers promised, was a tool that would outpace this kind of evolutionary change, something that would help breeders stay one step ahead of pathogens and pests. As the Brookhaven biologist Harold Smith summarized, making reference not to chemical-resistant insects but to the vulnerability of genetically uniform fields, "It may even be necessary to speed up the controlled evolution of organisms vital to our existence in view of the rapid alterations that humans are causing. . . . Consider, for example, the increasing menace from pathogenic organisms attacking crop plants when relatively homozygous genotypes, as of wheat, are grown over large areas."[13] Smith and others marshaled the trope of "speeding up evolution" not as a shorthand to explain their method but as evidence for why using nuclear technologies to induce mutation was a process ideally suited to meet the challenges created by modern, chemical-laden, monocropped production of highly inbred crops.[14]

A more general argument levied in support of mutation breeding took into account world population growth. This had already been used as a justification for the American research programs. Singleton, who had been a corn breeder for decades, knew that there was little domestic need to amplify corn yields. Nonetheless he argued that "in spite of these [already very high] yields, and the farm surpluses growing like the A bomb mushroom, it still behooves the scientist to learn all he can about efficiency of production and how higher yielding strains may be secured." His reasoning? Global population growth would force scientists to find means to achieve ever-greater production in order to feed a hungry world.[15] In the context of the international conference at Geneva, the argument for increasing global food production carried still more weight. The central statement on atomic energy and agriculture, attributed in the proceedings to the FAO but written by Silow and a representative of the USDA, focused on just this—the "world food problem" in relation to global population growth.[16] The message was not lost on attendees. One reporter present at the Geneva conference described how "nuclear physicists can help plant breeders to extend the frontiers of agriculture into the Arctic, into deserts and other regions where food crops have so far failed to grow" with obvious benefit to "a food-short world."[17] Or as another reporter concluded, "More food for the hungry world is the first prospect from such mutations."[18]

"More food for the hungry world" was a concern on the minds of many in the mid-1950s. A significant program that aimed to address just this problem was already underway in Mexico, and would become one of the best-known and most influential agricultural development programs ever undertaken. The initiative, begun in the early 1940s and funded by the Rockefeller Foundation, aimed to improve wheat and maize production in particular. Its successes, especially in producing high-yielding, disease-resistant wheat varieties and encouraging the adoption of intensive, industrial-style agricultural production, would later be credited with sparking the Green Revolution, first in Mexico and Latin America, and later in southeast Asia. These programs were closely tied to concerns about not only hunger but also the relationships among population growth, food security, and national and international security—concerns on the minds of governments, international organizations, and foundations alike in the early postwar (and Cold War) decades.[19]

As these examples indicate, the expansion of mutation breeding benefited from increased global attention to the present and future challenges of agriculture and plant breeding. Mutation breeding was not just a component of nuclear technopolitics but also global agropolitics. Following World War II, governments of developed and developing nations alike were eager to participate in efforts that promised to increase food production and in doing so enhance national security.[20] Breeders were equally eager to take advantage of new institutional resources and to try new technologies. With events such as the Geneva conference drawing attention to the perceived benefits of incorporating mutation techniques into plant breeding programs, mutation breeding captured the attention of many more researchers and government institutions. This included especially—though not exclusively—techniques that deployed nuclear technologies. A number of programs dedicated to the use of atomic radiation in plant breeding sprang up around the globe in the late 1950s and 1960s, taking their place alongside both new atomic and new agricultural research institutions constructed at this time.[21]

The Brookhaven program was a key player in this global dispersal of mutation breeding techniques and technologies.[22] This was in part a consequence of the expanded audience the Brookhaven researchers enjoyed while at the Geneva conference, which sparked demand from abroad for the services they offered. Their most extensive network of collaboration was among Canadian researchers, but countries represented in the pool of researchers aided by the Brookhaven program included both richer and poorer, American allies alongside former enemies, neighbors and distant states.[23] These collaborations typically centered on the improvement of specific crops. According to one report, the researchers from Italy hoped that irradiation might lead to an early-ripening almond, those from Thailand for a rubber plant resistant to specific diseases, and the Pakistani group for a taller jute plant.[24] Some collaborations had a more obvious political edge. In 1957, Brookhaven employees irradiated peanuts and soybeans for Chinese Nationalists in Taiwan, who reportedly hoped for "bigger producing, early maturing, disease resistant nuts."[25] Rather than sit back and wait for further interest, the Brookhaven program actively solicited it, notifying various foreign embassies of the program and of its willingness to assist researchers.[26]

Many institutions chose not to rely on Brookhaven for the use of irradiation facilities, and instead established their own independent research programs in mutation breeding. The AEC had supported the spread of gamma-irradiation technologies for agricultural research at several US institutions in the late 1950s, but its first foray in supporting the development of a gamma field abroad took place in Turrialba, Costa Rica, at the Tropical Center for Research and Graduate Training of the Inter-American Institute of Agricultural Sciences. The mission of this institute was to enhance agricultural education across Latin America; its contract with the AEC, signed in 1957, incorporated the application of nuclear science into this established educational mission. As such, it merged the atomic development agenda with that of agricultural development, "encourag[ing] the use of nuclear energy in solving problems relating to agriculture" across the region. Beginning in 1958, graduate students and other trainees at the Tropical Center had access to courses and research opportunities in the use of radioisotopes, whether used as tracers in soil or botanical studies or as a source for producing mutations in their six-acre gamma field.[27] The research undertaken with these tools centered on crops critical to agriculture in Latin America, including coffee, cacao, rubber, bananas, and palms. One of the first major studies to be undertaken in the gamma field (which had a 200-curie source of cesium-137) was an investigation of coffee genetics, a project in which it was hoped that the gamma field would produce mutations useful in both genetic research and the coffee breeding program at the institute.[28]

The Comitato Nazionale per l'Energia Nucleare (CNEN) in Italy reaped a similar harvest from the AEC's incorporation of gamma field development into its Atoms for Peace outreach. The CNEN gamma field went into operation in May 1960 at the Centro Studi Nucleari, about twenty miles north of Rome, with a radiation device that had been, according to a report of station biologists, "made available by the U.S. Government for the Atoms-For-Peace Programme." The CNEN researchers used the 250-curie source to study the radiation tolerance of common flowers, crops, and trees, including clover, cheatgrass, bellflowers, poppies, flax, tobacco, apple and pear trees, olives, grapevines, and durum and other Italian bread wheats. By 1962, the gamma field and associated research activities at CNEN appeared to be expanding in precisely the direction the Brookhaven program had a decade earlier. CNEN invited researchers from outside the institute to use the gamma field in their investigations or, alternatively, to have plants and seeds irradiated for them by staff members via an "irradiation service of the Laboratory of Plant Genetics."[29]

Other would-be recipients of Atoms for Peace largesse in the form of gamma field development and irradiation devices proved more independent. In 1957, the AEC approached the Japanese Atomic Energy Commission with a proposal: it would provide a 200-curie radiation device to support the construction of a gamma field for use by Japanese scientists. The offer prompted a review by the Japanese Ministry of Agriculture and Forestry, which decided that although a gamma field would be a useful research tool, the type offered by the US government (a "Brookhaven type irradiation device and cobalt source") produced too great a quantity of scattered radiation. Instead, "the Gamma Field Organizing Committee . . . decided not to accept the free-of-cost American device, but decided to construct a special irradiation device of her own and to purchase a cobalt source better suited to the special needs [of the program]," according to the report of a scientist from the newly developed Institute of Radiation Breeding. The new equipment, up and running by March 1961 with a 2,000-curie source purchased from Oak Ridge National Laboratory—a far more radiant source than had been offered by the AEC—was used primarily on agricultural and industrial plants, as well as forest trees and horticultural varieties.[30]

The list of sites hosting gamma fields by the mid-1960s included several in the United States, as well as Argentina, Costa Rica, Denmark, India, Italy, Norway, Spain, Sweden, Japan, the Philippines, and Puerto Rico.[31] Not that a catalog of the countries and institutions that had built gamma fields captures the full range or content of mutation breeding programs around the world in the 1960s: the fields were only one of many potential components of an induced-mutation breeding program, and they were not even the most effective of these tools. The extent of gamma field development was, however, a sign of how the Brookhaven mutations effort had created a blueprint for several subsequent large-scale mutation breeding programs (fig. 28). It provided the AEC and later the IAEA with a format and suite of technologies that these organizations would continue to promote—and, in the estimation of some, would promote without regard for their scientific foundations or indeed their benefits in comparison with other methods of plant breeding.

The proliferation of negative assessments of mutation breeding, even in the midst of its international expansion, presents strong evidence for the central role of technopolitics in its stabilization as a viable option for a plant improvement program. These criticisms came, on the one hand, in the form of close examination of studies that had been hastily conducted or performed

FIGURE 28. The Brookhaven gamma field, seen here from above, became the model for other gamma fields constructed in the United States and abroad. NARA RG 326-G, Box 6, AEC 57-5608.

without proper controls. Some of the more celebrated of the Brookhaven studies came under scrutiny, and further research demonstrated that they suffered from various experimental errors. This was true of the rust-resistant oat variety reported by Konzak in the early 1950s. By the end of the decade, the geneticist and breeder Richard Caldecott had shown that its rust resistance was not the result of an induced mutation but rather the product of increased hybridization among irradiated plants.[32] The apparent changes in carnation flowers were also demonstrated to be not mutations but alterations in the morphology of the plants resulting from the radiation treatment.[33] The reassessments seemed to bear out concerns that some of the research on mutation had been carried out without a basic knowledge of mutation genetics or the controls required for an induced-mutation experiment.[34]

Some researchers in the 1950s also expressed concerns about what they perceived as unwarranted enthusiasm for mutation studies. W. M. Myers, a plant breeder and geneticist from the University of Minnesota who was involved in mutation breeding work of his own, described induced-mutation studies as having "spread like a case of measles throughout our plant breeding programs." He worried that the attention did not arise in all cases from knowledge of or commitment to mutation methods.[35] Still others believed that the claims made about what mutation breeding could and would produce involved gross exaggerations, which in turn spurred uncritical application of the techniques. After the Geneva presentations, some geneticists were up in arms, finding the media coverage of the panels on radiation and genetics as having been, according to one report, "altogether too unbridled." As the biologist and science writer William Dick described, "Some of those reports went too far in conveying the impression that tremendous strides had already been taken bringing us close to the objective of refashioning the chromosomes of crop plants at will."[36] As a result of such criticisms and concerns, a few of the most ardent atomic-age mutation enthusiasts found themselves scaling back their statements. Even Singleton incorporated the errors in research at Brookhaven into a cautionary tale about what could and could not be reasonably expected of mutation breeding.[37]

Mutation breeding continued to have strong support in the 1960s and later—even into the twenty-first century. This support came most notably through a joint program on nuclear techniques in food and agriculture established in 1964 between the FAO and the IAEA. The FAO/IAEA program conducted conferences, coordinated research, and promoted awareness of mutation methods and new varieties produced through them, among other activities. The organization quickly established itself as the foremost institutional base for mutation breeding worldwide.[38] In time, a few success stories would be advanced to justify continued attention to mutation breeding programs. Most notably mutation techniques were said to contribute to the development of varieties celebrated as part of the Green Revolution. As the IAEA glowingly reported in 1969, "Mutation breeding had made a great contribution to the breakthrough called the Green Revolution, which had alleviated the problem of providing enough high-quality food to keep pace with the population growth."[39] There is reason to doubt the extent to which this was truly the case. Statements such as these exaggerated the influence of induced-mutation varieties and in fact were the source of much anger and conflict.[40] To those who might dispute the effectiveness of mutation breeding, the IAEA had an answer: even if other methods of breeding were more central to increases in agricultural productivity, mutation techniques nonetheless "had drawn attention to the significance of plant breeding and the number of characteristics of high-yielding varieties which needed improvement."[41] They had of course also drawn attention to the IAEA as a potential contributor to global food production as much as to other areas of nuclear development.

The creation of an international system for the development and promotion of atomic energy was crucial to the continued development of mutation breeding, even after decades of disappointment (dating back to the first efforts to use x-rays for breeding) and in the face of continued skepticism on the part of some practical breeders and research geneticists. The support of the AEC and later the IAEA, and indeed many other national atomic energy agencies, transformed mutation breeding from an uncertain and unproven set of technologies into an international activity with significant institutional support and extensive intellectual engagement across national borders.[42] In short, it transformed the field into an established—if sometimes controversial—area of research and development in plant innovation, one in which many researchers in many countries aimed to hone induced mutation as a precise tool for genetic modification.

Perhaps the most striking aspect of the resurgence of interest in the use of induced mutations for crop improvement in the postwar years is that it did not arise from the revelation that radiation derived from radioisotopes could be used to produce some effect on genes or chromosomes that had been long hoped for but consistently unattainable. The initial use of x-rays and colchicine in plant breeding had followed on assumptions, shared among many researchers, that agriculturists in particular would benefit from the development of a tool that would produce specific genetic changes (i.e., induced mutation or chromosome duplication). When the technology for producing these changes appeared, researchers sought to test their assumptions. Radioisotopes, however, were not new tools, nor were there well-established ideas about their potential usefulness to breeders when they were introduced into American agricultural research after the war. Changes caused by exposure to radium had been demonstrated to be quite similar to those resulting from x-rays, and radiation-induced changes had for the most part been deemed useful in genetics research but not in agricultural application.

It took the shared impulse of scientists and politicians to develop applications for new nuclear technologies, and the resources deployed to this end by the US government through the AEC, to create a shift in these attitudes in the United States. The Brookhaven research program proved key to the transition. There researchers such as Singleton moved from a sometimes skeptical renewal of induced-mutation research, to the exploration of its use in plant improvement, to an all-out embrace of mutation breeding and promotion of the idea, first in the United States and then around the world. The establishment of mutation breeding programs such as that of the UT-AEC Agricultural Research Laboratory, and the participation of many breeders in their work, confirms that a more positive assessment of radiation-induced mutation was quickly established, even as clear results—in the form of improved varieties—remained slow in coming.

It was not all public relations. Just as researchers who worked with x-rays and later colchicine characterized their work as an effort to craft effective and efficient tools for engineering plants to better meet the demands of agricultural production, so too did many who worked with radiation in the atomic age characterize themselves as perfecting a new technique. Some recognized that the resurgence of interest in using radiation in breeding was not primarily driven by new discoveries about its usefulness, but rather by the sharp rise in access to radioactive materials and nuclear technologies.[43] Far from undermining their agenda, this seemed to support the argument that it had only recently become possible to perfect radiation as a tool of the plant breeder: only now were many potent forms of radiation treatment more widely available, and only now had a whole coterie of researchers (many with AEC funding) taken up the study of radiation-induced mutation. Researchers set out to chart in detail the effects of modifying environmental factors such as water content or oxygen exposure, and to produce knowledge about the doses of radiation required to cause genetic changes in the wide range of species in which breeders might be interested. In short, they focused on ways of making gamma-ray emitters and neutron exposure facilities more reliable tools.[44] Better knowledge of the mutation process would take breeders closer to what the biologist Harold Smith of Brookhaven characterized as a communal goal: "We seek to control more effectively and to speed up appreciably the tailoring of useful plants and animals to meet our needs."[45] This ambition, and the rhetoric of more precisely manipulating plants, characterized many areas of mutation breeding research after 1955.

More important, of course, was the expanding technological system—with all its material, social, and intellectual components—that encouraged the development and use of nuclear technologies in the postwar decades. That, after all, is what rewarded mutation breeders (as they well knew) for their stated interest in making nuclear technologies into precision tools of genetic manipulation. The institutional support given to mutation breeding by nuclear agencies, especially the IAEA but also various national sponsors of nuclear research, proved effective in creating a self-conscious global community of such breeders from the 1960s onward. They marshaled whatever successes of induced mutation could be found in support of their continued research. These programs remained controversial, as did the basic notion of improvement via induced mutation. To critics, "speeding up" the appearance of a single variation was a process that involved the creation of many thousands of genetically damaged and useless plants, a slow and hit-or-miss process that required expensive infrastructure. They pointed out that in many countries improvement in basic plant breeding methods would be a more cost-effective investment toward enhanced food security than construction of a gamma field or other irradiation facility.[46] And of course there was still the paramount need for breeders, regardless of their national or institutional home, to create stable, uniform, predictable agronomic varieties. Creating random changes in the genetic constitution of a particular plant through heavy doses of radiation was at all times counter to that overarching goal.

It is tempting to decide that the efflorescence of mutation breeding, and its persistence, was all politics and posturing on the part of those who wanted either to promote nuclear development (the IAEA) or those who sought to gain professionally by continued support of and attention to their area of research (professional mutation breeders). But it is also important to recognize that these researchers followed in a tradition, dating to the turn of the century, of declaring induced mutation a promising method of altering genes and chromosomes to the ultimate benefit of agricultural production. As Singleton grandly claimed of mutation breeding in 1957, echoing those who had celebrated x-ray breeding three decades earlier, "No longer is the [plant breeder] limited to the mutations that occur naturally; he can make them occur. Whereas a particular mutant might occur only once in a hundred years, radiation can increase the frequency as much as a hundredfold, making it possible to produce desired mutants perhaps in one year. Hence evolution can be accelerated."[47] The crucial difference was that Singleton and his peers in the atomic age had greater access to resources than their predecessors. Whereas the failure of geneticists and breeders to produce anything particularly striking in the first decade or so of x-ray treatment contributed to the abandonment of this technology in American plant breeding, those using nuclear technologies found themselves buoyed by the growing atomic infrastructure even in the absence of clear achievements.

It was the radically different institutional and ideological implications of success in the atomic age—its being embedded in large technological systems and in technopolitical regimes in which atomic technology *had* to produce a variety of acceptable public goods in addition to atomic weapons—that made mutation breeding a robust area of research after World War II. If General Electric's interest in seeing x-ray mutation succeed as an agricultural method was great, the IAEA's interest in seeing atomic-aided breeding succeed was greater still. It directed resources toward producing, tracking, and cataloging types derived from mutation methods as evidence that would justify the continuation of mutation breeding activities and "peaceful atom" programming more generally.[48]

It is important to remember, however, that the appeal of these technologies was not limited to national and international sponsors and those researchers who benefited from interest in the promotion of atomic technologies. The

atomic gardening phenomenon in particular serves as a reminder of how enticing the idea of atomic-assisted evolution was beyond the walls of nuclear
and agricultural laboratories. As in the case of x-rays and colchicine, amateurs
expressed curiosity about novel technologies, both nuclear and genetic, and
showed an evident desire to wield them or at least to see them at work first-
hand. The content of one's garden could be rendered entirely unique, for as
one sales pitch put it, "it all depends on what strange atomic genetic reshuffling
took place during the special atomic processing."[49] Regardless of the debates
taking place about the effects of fallout from atomic testing or the other potential long-term hazards of radiation exposure on humans, in the garden "genetic reshuffling" via these promising atomic technologies had a clear appeal.

In the early 1960s, Singleton railed against the sale of irradiated seeds to
the public. He worried about the effect of hucksters who would make a quick
buck off an idea that he had helped popularize and in the process possibly discredit professionals like himself. But he missed the mark. It was not individual
hucksters like Speas, but the spinning and churning of the whole technological
system that produced atomic-energized seeds—just as it had also generated
Singleton's research program at Brookhaven and the whole gamut of mutation
breeding installations around the world. The imperatives of plant innovation,
felt as urgently as ever, had become tied up in the exigencies of atomic innovation. They have remained so since.

When it was demonstrated conclusively in the 1920s that x-rays and radium could be used to produce heritable variations in experimental organisms like *Drosophila* flies and maize plants, it was a given that these would soon be tested on the whole range of agricultural organisms. After all, the aspirations for a technology that could be used to produce such variations had been tied up with agricultural hopes from the start. Yet to gloss this simply as a moment in the history of American technoscientific production, in which it is impossible to distinguish something like applied outcomes from a prior body of scientific research, would be to miss the more striking lessons found when looking closely at this history, as well as at the history of colchicine and radioisotopes as plant breeders' tools. In each of these cases, breeders and experimenters tackled the particular challenges of innovating living products—here, breeding new crop plants and garden flowers—via modes more commonly associated with the innovation of nonliving things. Yes, plant breeding was technoscience. As such, it also shared in the diverse forms of twentieth-century technoscientific production, whether directed industrial research, open-ended tinkering, or expansive system building.

As I argued in part 1, the pursuit of x-rays as a technology of genetic manipulation and plant innovation is best understood as an effort to industrialize biological innovation. In the early decades of the twentieth century, many American geneticists, breeders, and others linked the controlled and efficient production of genetic mutations—believed to be the source of the heritable variations seen in nature—to the controlled and efficient production of new crop varieties. If x-rays could be made to produce mutations in this way, then plant innovation might be made to more closely resemble other areas of technological innovation. The latter had already been rationalized, folded into industrial production via the creation of industrial research laboratories. If the same could be accomplished in plant breeding, not least by "speeding up evolution" with x-rays, it would better align floundering American agricultural producers with their flourishing industrial counterparts.

In part 2, I made a case for the important place of tinkering—and tinkerers—in the history of plant innovation. After years of experimenting with various unreliable techniques that on occasion produced desirable chromosomal changes, breeders began in 1937 to use the chemical colchicine, which appeared to consistently induce the doubling of chromosomes in many plant species. Like x-rays, colchicine promised a means of controlling evolution. It apparently enabled direct and precise manipulation of the genetic material, in ways that mimicked changes also known to occur naturally. Most professional breeders saw in this chemical a means of distancing themselves from older, less reliable approaches—from not only the need to continuously tinker with different techniques of manipulating chromosomes but also the guessing game carried out with genes when hybridizing plants or applying mutagens like x-rays. Not every breeder was concerned solely with precision and efficiency, however. Amateurs proved eager to experiment with the chemical, often wanting simply to experiment with breeding methods and to tinker with genes and chromosomes on their own. Some commercial flower breeders similarly embraced the use of colchicine as a generator of random change (precisely the kind of thing that colchicine's original promoters had hoped to move away from), seeing it as a quick and appealing route to producing flower novelties. These tinkering technologists, amateurs and professionals alike, ensured that the use of colchicine in plant innovation would continue even when it fell short of the promised goals of precision and efficiency.

After World War II, attention turned back to radiation as a potential tool of plant innovation. This time, however, the radiation sources were not x-ray tubes but novel atomic technologies such as nuclear reactors and massively radiant isotopes. I argued in part 3 that these atomic-age innovations in breeding saw extensive use not because they promised to solve pressing problems in plant breeding but because they furthered the aims of politicians, satisfied the ambitions of growing scientific institutions, and further expanded entrenched infrastructures. Plant breeding innovations that relied on atomic energy emerged as components of a large technological system dedicated to the development of atomic energy in the United States. Like many innovations produced within such a system, they did not upset the established order but instead extended it into new domains. The use of these technologies was not limited to the system that generated them, however, for as they became increasingly popular, they were incorporated into another large technological system, that of modern industrial agriculture. The result was further conservatism in innovation, in which the production of innovations in breeding technologies was geared to the further extension of an atomic infrastructure, and the production of plant innovations (i.e., new breeds) was directed to the further entrenchment of industrial agriculture. This was true in the United States in the 1950s, but even more so around the world after 1960. The near simultaneous internationalization of the atomic infrastructure and globalization of industrial agriculture proved crucial to the long-term survival of the atomic-inspired innovations in plant breeding and of an approach known today simply as mutation breeding.

These stories highlight the continuity of ideas about and approaches to innovation across varied technoscientific fields and diverse products in the twentieth century; they also highlight the temporal continuity in technological ambitions for genetics and plant breeding. Throughout these decades, the same expectations accompanied each new innovation in genetic technology. Each was understood to provide a means of producing heritable variation on demand, thereby speeding up the rate of evolution in almost any plant species. This in turn would enable breeders to innovate new and better varieties more efficiently, as they would have access to a greater range of variation as and when they needed it. The imagined end product was not just a series of improved crops: it was a transformed mode of plant innovation in which slow natural processes of evolutionary change would be supplanted by rapid human-controlled genetic alteration.

These expectations proved far too great. At each iteration, new techniques failed to make plant breeding any faster or more efficient than it had been before. They did not make the production of variation a routine or reliable process. But unrealized expectations did not equal failure, nor did they entail abandonment of these tools. Many users harbored expectations for these genetic technologies that extended beyond efficient crop improvement. A technique ideal for a tinkerer's toolkit (like colchicine) or a technology useful in international cooperation (like cobalt-60) could continue in use, regardless of its proven capacity for generating useful heritable variations and therefore new varieties for the market. Nor did the successive failures of these technologies to transform plant breeding dash hopes for still other tools that would be transformative. Anticipation of such technologies and the future they would make possible persisted—in the minds of experimenters, in the business models of commercial producers, and on the pages of newspapers.

Albert Blakeslee used the term "genetics engineer" to describe his methods of creating "new species to order" in 1937. When the close cousin of this term—"genetic engineering"—appeared again in the American lexicon, this time to stay, it was used in reference not only to plants but also to people. "And Now Your Child Built to Order!" declared the headline of a 1965 article in the *Chicago Tribune* about the advances made by biologists in manipulating DNA. With the "day of [the] genetic engineer near," scientists such as the prominent biochemist Rollin Hotchkiss encouraged the public to become more knowledgeable about the new capabilities of biological research. "The potential for genetic modification of man by artificial intervention is with us now," Hotchkiss reportedly declared.[1] The genetic technologies under discussion were those that had emerged from the coalescing discipline of molecular biology in which biologists were turning their attention to the manipulation of bacterial and viral DNA through various means, and claiming its eventual (and rapid) extension to human genetic material.[2]

Assessments of molecular biological research in the late 1960s bore many similarities to earlier speculations about the use of radiation and chemicals to modify genes and control evolution. A *New York Times* piece described the work of molecular biologists as "evolution control," touting technological advances that offered the "distinct possibilities of redirecting the entire course of human and animal evolution." The specific findings discussed seemed as yet distant from application, such as the work of Jacques Monod and François Jacob on the genetic regulation of protein production in *E. coli* bacteria or the synthesis of viral DNA by the biochemist Arthur Kornberg, but this did little to temper speculation. The biologist Bentley Glass confirmed that molecular biological tools truly offered new capabilities to human beings. "I don't know how soon man will be willing to use the power . . . but we have the power now to alter evolution," he told the *Times*.[3]

In some cases, discussion of the new genetic engineering prompted a celebration of modern science and technology. Many news reports of molecular biologists' achievements conveyed hope and wonder about a future in which human health and well-being would be improved through new therapeutics. Humans might even be pushed beyond their known capacities. One reporter speculated on the use of "genetic surgery" to enhance athletes' physical prowess, writing, "Sporting events would become battles among geneticists, making a mockery out of contests such as the Olympics."[4] But the overwhelming focus on the application of genetic engineering to humans, as opposed to plants or even other animals, also led to reflection on the ethical issues such engineering might raise. Who would decide which traits were "good" and which were "bad" in breeding better human babies, for example? Scientists and nonscientists alike pondered whether it was appropriate to "tinker" or "tamper" with human evolution in the ways that quite a few researchers were proposing would be possible.[5] This concern had rarely accompanied earlier claims to controlling the evolution of plants with x-rays or chemicals.

The atmosphere was charged with these ideas and their implications when the Harvard biologist Jonathan Beckwith and his research team announced in 1970 that they had successfully isolated a gene (that is, extracted the segment of DNA that coded for a particular enzyme) from a bacterial chromosome—and at the same time released a public statement about the dangers they foresaw in genetic engineering.[6] One member of the research team described to the *Boston Globe* such things as placing "viruses containing a sterility gene" into water supplies to reduce unwanted populations and "eliminat[ing] dissent by injecting genes into humans which would make their behavior more placid." As he summed up, "Changing the human race via genetic engineering is no small thing."[7] Biological research, genetic technologies, evolution to order—these were things that carried the potential for good, yes, but they were also a cause for concern. For many, such sentiments only intensified with the development of in vitro fertilization and other human reproductive technologies in the late 1970s, given the seeming potency of combined genetic and reproductive control.[8]

Over the course of the 1970s, as molecular biological research led to more innovations in genetic engineering and more laboratories took up research along these lines, concerns about safety took their place alongside ethical questions. A team of researchers led by the biochemist Paul Berg first succeeded in the transfer and expression of genes from one species in a second in 1974, with an experiment that incorporated viral DNA into a bacterial genome. Immediately the thought arose among these researchers that such a technique could lead to either the inadvertent or the intentional creation of a new human pathogen. Led by Berg, a group of concerned biologists called for a voluntary moratorium on recombinant DNA research, and subsequently they organized a conference to debate among themselves how such research could best be governed to ensure safety.[9] Concerns among scientists—and their eventual decision to restrict oversight of recombinant DNA research to members of the scientific community—spun into popular debate and in some cases public hearings. Alfred Vellucci, the mayor of Cambridge, Massachusetts, cited concern for public welfare in light of scientific hubris in one such hearing. As he told one reporter, "It is my responsibility to investigate the danger of infections to humans. They may come up with a disease that can't be cured—even a monster. Is this the answer to Dr. Frankenstein's dream?"[10]

It took biologists almost another decade to develop recombinant DNA techniques—the transgenic methods now associated with the notion of genetic engineering—that could be used in plants.[11] When they did emerge (and even before), researchers and journalists were quick to tout the potential benefits that would arise from molecular manipulation of plant genomes. These included enhanced nitrogen fixation, improved nutritional quality, greater tolerance of drought and cold, and better disease and pest resistance. One *Washington Post* journalist, who dubbed these transgenic plant experiments the "test-tube babies of agriculture," described the ways in which breeding would change forever once the transgenic methods had been perfected: "Classical methods can take 10 years to produce a new hybrid. . . . But, by manipulating genes in the lab, researchers can make new plants more quickly and precisely." And of course, there would also be the possibility of carrying genes "across all biological boundaries," an option previously unavailable to the plant breeder.[12]

These claims to speed and precision echoed the claims made by those who earlier had described mutagenic techniques as means to manipulate plant genes and master evolution. But transgenic methods of plant breeding were not understood by comparison to mutagenics, though the latter continued to be used by breeders in many parts of the world. Transgenic techniques were one part of the touted-to-be-revolutionary toolkit of molecular biology. As such, they were discussed primarily in relation to the uses of molecular biology to address human medical and social concerns. This was a context in which these innovations in plant breeding gathered ever more attention from the media and in the marketplace, as well as from geneticists and breeders at commercial and nonprofit institutions around the world.

It was also a context in which the perceived desirability of making life "to order" had shifted dramatically. Those who would create transgenically altered plants faced some of the same objections to the outcomes of "tinkering" and "tampering" that other products (real and imagined) of the new genetic technologies had produced. Perhaps most significant were concerns surrounding the "release" of transgenic plants via their use in agriculture as crop plants. Some worried that transgenic crops grown out in open fields might become aggressive weeds that, like the imagined pathogens of the transgenic bacterial experiments, could not be contained. Other observers speculated on the potential for other forms of ecological damage, such as harm to native insect populations from genetically altered pollen or the creation of ever-more-resistant pests and pathogens as a result of increased genetic homogeneity.[13]

Such concerns captured national attention in the United States in 1983 when activists filed suit against the National Institutes of Health and the University of California over a planned field trial of a transgenic organism. The organism in question was not a crop plant but an engineered bacterium. Scientists hoped its unique properties would prevent frost damage when sprayed on crops. Activists succeeded in having the field test of these "ice-minus" bacteria delayed, raising enough local opposition and negative press attention to prevent trials for several years. As one report summed up, "Opponents argue that, although they may save time, genetic engineers could cause nature to lose its balance by releasing into the environment new organisms, the effects of which are unknown."[14] Clearly the perceived implications of intervening in natural evolutionary processes had changed. For many, speeding up or altering evolution now posed threats.[15] In the context of agricultural production, it posed threats in particular to the natural environment, but it also created worries about human health effects that might arise from the genetic alteration of food crops.

Of course, many aspects of American society had changed in the two decades between the fad of sowing atomic-energized seeds in the early 1960s and the concerned response to transgenic plants in the 1980s. Perhaps most important for this story, the environmental movement had burst into American culture and politics, providing new sensibilities about human effects on and responsibilities toward the natural world.[16] Also important were the rights movements of the 1960s and 1970s, the protests of the Vietnam War, and the questioning of both governmental and scientific authority that accompanied these.[17] And there was the growth of the biotech industry, too, a sector strongly supported by the federal government. Industry interest in agricultural products aggravated existing concerns about the corporate control of this market—for industrialization and corporatization in agriculture had become hot-button topics in both domestic and international politics. In this context, objections to genetically modified crops took myriad forms, and human intervention into natural evolutionary processes was, though common, only one of many.[18]

The implementation of early genetic technologies such as x-rays and chemicals differed significantly from that of later innovations such as transgenic techniques. Most obviously, the latter were far more successful in their being used to develop breeds that saw widespread use in agricultural production. Mutagenic methods, though still in use today, rarely achieved anything as sweeping as was claimed for them. The most famous products of mutation breeding in the United States are one-off successes in various crops, such as a couple of popular red grapefruits, Star Ruby and Rio Red, and a variety of short-stature rice called Calrose 76.[19] There are many varieties known to have been produced through mutation breeding that are grown in the United States and around the world, but even a glance at the acres planted in crops created via transgenic technologies is enough to show that the more recent innovation had a far greater effect on agriculture. In the United States in 2015, transgenic crops accounted for 94 percent of acres planted in soybeans and cotton and 92 percent of acres planted in corn.[20] There are no equivalent data for mutagenic varieties but obviously the cultivation of these is insignificant by comparison.

This straightforward measure of productivity is important, but in some ways it is the least interesting comparison that can be made between older and newer genetic technologies. More thought-provoking distinctions emerge when one applies some of the same approaches to the history of innovation used in this book to explore different aspects of the innovation of transgenic technologies and to compare these with earlier innovations: What were the dominant visions of what genetic technologies might produce or enable at the time these technologies were initially created? What were the aims of those who used these technologies and how did that influence their subsequent development? Were transgenic innovations produced within a larger technological system and, if so, what role did they play within it?

The answer to the first of these questions is partially answered in the discussion above. Innovations that would enable transgenic manipulation for plant breeding were first discussed and then introduced in a context in which ideas about the potential implications of genetic manipulation were far more mixed than they had ever been before. From the 1960s onward, Americans were increasingly aware of the social and environmental costs of unfettered scientific and technological development, from nuclear power to Cold War missile defenses to chemical-intensive agricultural production. Many more people—including scientists and engineers who wished to distance themselves from the criticisms made of the atomic bomb scientists—were ready to question the benefits of new technologies and to assess their potential effects before allowing something potentially harmful to be launched into the world.[21] So when molecular biologists claimed, far in advance of their having the ability, that their tools would enable the genetic engineering of living organisms in heretofore-unimaginable ways, one response was not to celebrate but to interrogate, and even to fear. This was true not only in the imagination of human genetic manipulation, where the specter of eugenics loomed large. It was even true for the genetic manipulation of plants. There were possibilities for creating better crops, yes, but also opportunities for unwittingly producing ecological disasters.

Not all responses were critical, however. One important continuity between the history of early genetic technologies and that of more recent innovations in genetic manipulation is the extent to which researchers, aided by news reporters and corporate press offices, made claims for their technologies that far outstripped any proven capabilities. Consider the cases of x-rays, colchicine, and nuclear technologies: none of these innovations ever matched the extraordinary predictions set forward in the earliest stages of their development. And researchers who expressed skepticism were, by and large, ignored. It was hype, not realism, that fueled research careers, sold newspapers and magazines, and indicated that public investment in science did indeed produce results. The same continues today with more recent developments in biotechnology. One need only consider the significant financial speculation in biotechnologies in the 1980s, when investors poured money into biotechnology start-ups in response to the promises these firms made about the revolutionary potential of their research, and the disparity between these expectations and later outcomes.[22]

Comparing older and newer genetic technologies from the perspective of users and technological systems is also illuminating. Many of the professional researchers who worked in induced-mutation breeding were based at public institutions, working in the public interest. They were employees of state agricultural experiment stations, the United States Department of Agriculture, and national nuclear laboratories.[23] They had access to very little in the way of intellectual property protections in the living organisms with which they worked and few reasons to pursue such protections. Commercialization and control of any particular market for seeds or crops were simply not on their agendas. This was less true for the employees of General Electric—who tellingly patented their only plant production—and certainly not true of David Burpee who no doubt would have loved to have greater legal control over the dissemination of his flower innovations. Still, the contrast to today's configuration, in which transgenic crops are vigorously pursued by large multinationals that have significant control of the global market in seeds, as well as the ability to patent their GM varieties and the financial resources to enforce those patents, could hardly be more stark.[24]

It is this latter configuration that has led many observers to consider the development of transgenic crops as little more than a means of further entrenching a technological system that many find objectionable—that is, a high-input global agricultural system in which the production and sale of seed has been consolidated into the hands of a few powerful global corporations.[25] One need only point to the chemical company Monsanto, an early developer of transgenic techniques in plant breeding, for a clear example of these as innovations designed to entrench and extend the large technological system of modern industrial agriculture. The company is notorious for having introduced in the 1990s Roundup Ready transgenic crops, plants genetically engineered to be resistant to the herbicide Roundup, also a Monsanto product. The crops and chemical were thus a package deal, the use of one encouraging the adoption of the other.[26]

Even here, however, parallels may be drawn to earlier genetic technologies on closer inspection, particularly in the desire to deploy technologies because they promise "more" and "faster" without also reflecting on how either of those would improve the agricultural system as a whole. The appeal of using mutagens to produce novel types in the midst of the agricultural depression of the 1920s and 1930s made a certain kind of sense. Innovating new crops for the market might alleviate some of the problems created by the overproduction of staple crops. But other proposals, such as the use of radiation in the 1950s to "speed up" the evolution of crops so as to stay ahead of the evolution of insects and pathogens, are more difficult to understand as improvements to agricultural production as a whole. The problem of resistance intensified significantly in the postwar years as industrial, monocropped, chemical-intensive agriculture became ubiquitous.[27] In this case, the use of radiation was imagined as a quick technological fix to a problem created by industrial agriculture, and not a means of addressing the root of the issue by adjusting the system as a whole to be more sustainable. Breeders offered this innovation as a means to preserve the whole system intact, just as large companies initially used transgenic technologies in ways that conserved the status quo.

Of course, like radiation breeding, transgenic technologies are not in themselves limited to such system-conserving applications, nor are employees of multinational corporations the only plant innovators who put these to use. There are many researchers, including many employed at public institutions, who hope to use transgenic technologies in more radical ways. They envision these technologies as tools to create plants more amenable to smallholders or peasant farmers, for example, or plants that are more nutritious or that deliver vaccines. Some transfer their transgenic plants, or the knowledge behind their creation, directly to the public domain.[28] Although many organizations that oppose transgenic manipulation are not inclined to differentiate among various uses and users of GM crops, such differentiation does indeed matter, for projecting the future trajectory of a technology as much as for understanding its past.[29]

As I hope I have made clear, evidence of a broad acceptance of genetic technologies in the 1930s, 1940s, and 1950s does not mean that disapproval of the same in the twenty-first century is misguided. Rather, it indicates that many people have come to a different understanding of genetic technologies and the implications of their use in agriculture. This new understanding is informed by, among other things, a heightened awareness of the environmental consequences of most human activities, the entrenchment of industrialized agriculture, and the consolidation of seed ownership by a handful of multinational firms. Although the eventual benefits promised to arise through the use of genetic technologies remain very similar, these promises are accompanied by visions of harm and danger. And what continuities do exist between earlier discussions and uses of genetic technologies and those that occur today—the two most obvious being the hyping of technologies and the haste to put these to work in service of a technological system that is possibly destructive and at very least dysfunctional—should make one still more wary of claiming that the acceptance and even celebration of genetic technologies in an earlier era is any reason to reconsider more recent misgivings about the benefits these confer.

There is one final commonality to the many twentieth-century efforts at genetic manipulation, whether transgenic or mutagenic, which I hope this book has made abundantly obvious. From fascination with radium at the turn of the century to scientific and industrial interest in x-ray tubes, from faith in better living through chemistry to the embrace of atomic energy, researchers hoping to find tools for altering genes looked immediately to those technologies newly dominant elsewhere in American life. The evidence for this trend goes beyond the cases described here. In November 1948, a navy officer and an agronomist from the New Mexico Agricultural Experiment Station launched cotton seeds 90.5 miles into the stratosphere on a V-2 rocket, hoping to discover effects of natural stratospheric radiation on the genes of this species.[30] Studies of the potential for cosmic radiation to be used in inducing mutations in plants, whether on board manned spacecraft or in recoverable satellites, were conducted through the space age.[31] And the trend continues even today. With the biotechnologist J. Craig Venter's 2010 success in creating what he calls a synthetic organism, a bacterium with a genome that was designed by a team of researchers on a computer and synthesized in the laboratory, it is possible to add even computer and information technologies to the list of technologies deployed as the key to engineering life. In Venter's words, the product of his team's work is "the first self-replicating species we've had on the planet whose parent is a computer."[32] Whether this characterization of the research accurately captures the process is debatable, but Venter's desire to link his achievements with the dominant technologies of the twenty-first century is surely not.

Americans have long imagined the existence of a quick and easy route to innovating living things—especially, though not exclusively, the plants that underpin American agricultural production. That route has sometimes depended on other areas of innovation, such as industrial production, nuclear development, or computer engineering. It has at all times been indistinguishable from other areas of innovation, however special we may perceive its object to have been. Efforts to innovate tools and techniques that would allow Americans to innovate new forms of life did not constitute a specialized program of research, conducted by certain kinds of biologists in specific institutional settings. It was instead a generalized aspiration, something envisioned and pursued by many individuals, scientists and nonscientists alike, in many different settings, whether academic, industrial, commercial, or domestic.

Acknowledgments

There are many individuals and organizations to thank for the help and encouragement I received while researching and writing this book.

Thanks above all to Dan Kevles, who championed the project and its author from the outset and who continues to dispense advice and provide inspiration at each new juncture. His confidence in my work has made all the difference. Thanks to Bettyann Holtzmann Kevles, too, who was equally present and encouraging from day one.

Thanks to the other scholars who have mentored me or offered much-needed input along the way, especially Paola Bertucci, Janet Brown, Luis Campos, Soraya de Chadarevian, Karen Hébert, Bill Rankin, Paul Sabin, Jim Scott, and John Warner. Special thanks go to Angela Creager, not only for advice and support, but also for her generosity in sharing research materials related to the AEC and pointing me in the direction of useful texts and archives on a number of occasions.

Thanks to my colleagues in the Department of History and Philosophy of Science (HPS) at the University of Cambridge, where I brought this project to completion in as pleasant and stimulating an environment as one could ever ask for. Thanks in particular to Nick Hopwood, Simon Schaffer, and Jim Secord for reading the manuscript and alerting to me to many errors and opportunities I would otherwise have overlooked. Thanks also to the amazing students of HPS, who inspire me with their creativity, energy, and intelligence, and who are the reason I love my job.

Thanks to those who made writing a more social, and infinitely more enjoyable, endeavor—from my first-ever writing group of Francesca Ammon, Kathryn Gin, Julia Guarneri, Sara Hudson, Catherine McNeur, and Robin Morris, to the unparalleled sounding board of Henry Cowles, Joanna Radin,

Lukas Rieppel, Alistair Sponsel, and Laura Stark, to my far-flung writing buddies Jenny Bangham, Tiago Mata, Lisa Onaga, and Chitra Ramalingam.

Thanks to the many other scholars who shared ideas and offered critical feedback along the way, especially Hasok Chang, Courtney Fullilove, Jung Lee, Brendan Matz, Dmitry Myelnikov, Josh Nall, Jesse Olszynko-Gryn, Joy Rankin, Rachel Rothschild, Robin Scheffler, Kathryn Schoefert, Suman Seth, Richard Staley, Kristoffer Whitney, and Anna Zeide.

Thanks to the amazing teachers I've had over the years, especially George Repella, Larry Buell, David Spanagel, and the incomparable Glenn Adelson. They cultivated my love of reading and writing and encouraged me to think I might make a career out of these very activities. This was an incredible gift. Thanks also to Marty Malin who, though not in the capacity of a teacher, taught me a whole lot about working in the academic world and gave me my first opportunities to write and edit professionally.

Thanks to the libraries and, more important, the archivists and librarians, who helped me discover the materials needed to tell this story. These include Charles Greifenstein and his colleagues at the American Philosophical Society, Clare Clark at the Cold Spring Harbor Laboratory, Janice Goldblum at the National Academy of Sciences, Polly Armstrong of the Special Collections at Stanford University, and Mary Beth Brown and her colleagues at the Western Historical Manuscript Collection in Columbia, Missouri, as well as archivists at the National Archives and Records Administration in College Park, the Niels Bohr Library and Archives at the American Institute of Physics, the Bancroft Library of the University of California, Berkeley, the Special Collections Library at Clemson University, the Carl A. Kroch Library at Cornell University, the Lilly Library at Indiana University, Bloomington, the University Archives of the University of Missouri, Columbia, the Special Collections of the University of Tennessee, Knoxville, and the Small Special Collections Library at the University of Virginia. Thanks also to the amazingly patient and helpful staff of the US Library of Congress, who helped me track down the books, magazines, and newspapers from which many of the illustrations in this book were derived.

Thanks to the other institutions that made this project possible by providing resources to support research and writing, including the Agrarian Studies Program at Yale University, the American Council of Learned Societies and Andrew W. Mellon Foundation, and the Chemical Heritage Foundation.

Thanks to the University of Chicago Press for producing this book, especially Karen Merikangas Darling and Evan White for making the review and revision process painless, even pleasant, with their professionalism and attention to detail, and Kelly Finefrock-Creed for her careful copyediting. Thanks to my two reviewers for not only seeing promise in the original manuscript but also being ruthless in their critiques. I am particularly indebted to Audra Wolfe, who generously shared research materials and later went above and beyond in offering detailed feedback on the manuscript in two versions. This is a far better book for her attention to its smallest details and grandest aims alike.

What errors and omissions remain in this text are of course solely my responsibility.

Thanks to my grandparents, now deceased, Margaret and Edward Curry and Anna Marie and Francis Dougherty. Their hard work and sacrifices made it possible for a granddaughter to one day have big dreams.

Thanks to my brother and sister, Tom and Bridgid, for being there at peaks and troughs alike, usually with a cold beer and always with the same concern, intelligence, and sense of humor. Thanks to their spouses, Michelle and Amos, who have been no less generous and supportive, not least in housing and feeding me during many archival trips. And thanks to my nieces and nephews, Thomas, Anna, Elizabeth, and Charlie, who have done more than anyone to help me hone my storytelling abilities.

Thanks—endless and still insufficient thanks—to my parents, Tom and Anne Curry, who taught me all the things that really matter. They are wellsprings of love, patience, and generosity, which I have drawn on heavily. This is book is for them, in gratitude.

Thanks, finally, to Andrew, for all the adventures, especially those still to come.

Notes

Introduction

1. On American wheat breeding, see Olmstead and Rhode, *Creating Abundance*, ch. 2; on cytoplasmic male sterility in maize breeding, see Mangelsdorf, *Corn*, 239–40; for a technical account of genetically engineered tomatoes, see Hightower et al., "Expression of Antifreeze Proteins." Breeders have also relied on so-called bud variations, somatic mutations that appear in a bud or shoot of a growing plant that differentiate that part of the organism from the rest and that can be propagated via vegetative (asexual) reproduction.

2. Early, "Giant Flowers," 6.

3. Histories of technology in the United States include Cowan, *American Technology*; Hughes, *American Genesis*; Pursell, *Machine in America*. See also contributions to Pursell, *Companion to American Technology*.

4. On the industrialization of American agriculture in the twentieth century, see Gardner, *American Agriculture*; Fitzgerald, *Every Farm a Factory*; Conkin, *Revolution Down on the Farm*; Anderson, *Industrializing the Corn Belt*. See also Cronon, *Nature's Metropolis*; Danbom, *Born in the Country*.

5. Fitzgerald, "Technology and Agriculture."

6. On the American seed industry and horticultural markets, see Lyon-Jenness, "Planting a Seed"; Kevles, "Primer of A, B, Seeds."

7. Bashford and Levine, *Oxford Handbook*. Histories of eugenics in the United States include Kevles, *In the Name of Eugenics*; Paul, *Controlling Human Heredity*; Kline, *Building a Better Race*; Stern, *Eugenic Nation*.

8. On the long history of biotechnology (where this is taken to mean the use of living organisms in industrial applications), including fermentation and agricultural breeding, see Bud, *Uses of Life*; for further accounts of plant and animal breeding as a technological endeavor, see contributions to Schrepfer and Scranton, *Industrializing Organisms*. On landscaping and horticulture, see Pauly, *Fruits and Plains*. On experimental physiology, see Pauly, *Controlling Life*. On cell culture, see Landecker, *Culturing Life*. On genetics and molecular biology, see Kay, "Life as Technology"; Kay, *Molecular Vision*. Other scholars who engage with a long view of biotechnology include Creager, "Biotechnology and Blood"; Rasmussen, "Forgotten Promise"; Gaudillière, "New Wine."

9. Luis Campos similarly situates the development of mutagenic techniques among ambitions to control evolution or engineer life. See Campos, *Radium*; Campos, "Synthetic Biology."

10. "Evolution to Order," transcript of Science Service radio broadcast, 24 March 1938, Blakeslee Papers, Folder "Lecture, Papers, Etc.: Adventures in Science."

11. "Better Regal Lily," N1.

12. Throughout this book, I use the term "innovation" in preference to "invention" in light of the broader range of activities associated with the former in both scholarly and other use.

13. I discuss Burbank in chapter 1. For a celebratory biography of Borlaug, see Hesser, *Man Who Fed the World*.

14. E.g., Fitzgerald, *Business of Breeding*; Kloppenburg, *First the Seed*. Survey accounts that cover a wide range of innovations in technique (including some of those discussed here) are Murphy, *Plant Breeding*; Kingsbury, *Hybrid*.

15. This was a key insight in the development of new approaches to the study of technology beginning in the 1980s. See, e.g., Bijker, Hughes, and Pinch, *Social Construction of Technological Systems*; Bijker, *Of Bicycles*; MacKenzie and Wajcman, *Social Shaping of Technology*; Oudshoorn and Pinch, *How Users Matter*; Edgerton, *Shock of the Old*.

16. Lipartito, "Picturephone." On "failed innovations," see also contributions to *Social Studies of Science* 22, no. 2 (1992).

17. The classic survey of the history of biotechnology relying on this broad definition is Bud, *Uses of Life*. For other more-inclusive accounts of biotechnologies, see note 8.

18. See, e.g., Borlaug, "Ending World Hunger"; Prakash, "Crop Debate"; Fedoroff, "Past, Present and Future."

19. See, e.g., Morange, *Molecular Biology*, ch. 16; and the accounts of genetic engineering as found within the histories of industrial biotechnology listed in note 23. Even Lily Kay's strongly revisionist account of genetic technologies centers on molecular biology; see Kay, *Molecular Vision*. An account of plant-specific genetic technologies is Lurquin, *Green Phoenix*; for more popular accounts of plant genetic technologies, see Charles, *Lords of the Harvest*; Pringle, *Food, Inc.*

20. E.g., Campos, *Radium*; Campos, "Synthetic Biology."

21. The hopes pinned to genetic technologies in turn resembled those pinned to many different technological possibilities over the course of the twentieth century. The classic account of American "technological enthusiasm" is Hughes, *American Genesis*; see also Wright, *Possible Dreams*.

22. An exception among histories of biotechnology is Bud, *Uses of Life*.

23. On the history of the modern biotechnology industry, see Kenney, *Biotechnology*; Krimsky, *Biotechnics and Society*; Thackray, *Private Science*; Hughes, "Making Dollars"; Cooper, *Life as Surplus*; Hughes, *Genentech*; Rasmussen, *Gene Jockeys*. An account that stresses popular demand for useful products of life science research is Vettel, *Biotech*.

24. Kloppenburg, *First the Seed*; Schurman and Munro, *Future of Food*.

25. Several authors have surveyed responses to genetics in popular culture and public debate in the twentieth century. My account differs in two significant ways: it focuses on the uses of genetics research in agriculture and it emphasizes the interconnectedness of research activities and broader public engagement with these. See van Dijck, *Imagenation*; Turney, *Frankenstein's Footsteps*; Condit, *Meanings of the Gene*; Nelkin and Lindee, *DNA Mystique*.

26. Philip Pauly, for example, focuses on the vision of the biologist Jacques Loeb and his influence on later biologists; Lily Kay is principally concerned with the Rockefeller Foundation and its goals; Robert Bud explores industrialists and the scientists who support them. See note 8.

27. For studies that consider expectations in relation to biotechnologies, see Brown, "Hope against Hype"; Hedgecoe and Martin, "Drugs Don't Work." For an overview of research on expectations in science and technology, see Borup et al., "Sociology of Expectations."

28. It is increasingly common not only to present mutagenic techniques as an early precursor to contemporary genetic engineering (as I do here) but also to suggest that their unquestioned use over many decades in breeding food plants is a reason to be less concerned about or less questioning of transgenic manipulation. See, e.g., Fedoroff and Brown, *Mendel in the Kitchen*; Grun, "Defining the Term 'GM'" (including Fedoroff's "Response").

Part One

1. Tyler Archer, "The History of the National 4-H Camp, 1927–1956," accessed 9 June 2014, http://4-hhistorypreservation.com/History/4-H_Camp; see also Wayne Short, "My Trip to Washington" [scrapbook of 1928 participant], accessed 9 June 2014, http://4-hhistorypreservation .com/History/4-H_Camp/Wayne_Short_My_Trip_to_Washington_1928.pdf. For Slosson's address, see "Farm Boom Beginning."

2. Dennis, "Accounting for Research," 481.

3. E.g., Meyer-Thurow, "Industrialization of Invention."

Chapter One

1. Reported in Davenport, "First Report of Station" (quotation on 39).

2. Ibid., 43, 48 (quotation). On the early intersection of interest in radiation and mutation, see Campos, *Radium*, ch. 3.

3. On Mendelism and its reception, see Olby, *Origins*; Bowler, *Mendelian Revolution*. A useful overview of the historical debates about Mendel's experiments is Sapp, "Nine Lives."

4. For one discussion of the rediscovery, see Olby, *Origins*, ch. 6; see also the summary in Bowler, *Evolution*, 264–68. On the development of genetics in relation to Mendelism, see the references in note 5.

5. On chickens, see Cooke, "Science to Practice." On genetics and agriculture in the United States, see Kimmelman, "American Breeders' Association"; Kimmelman, "Progressive Era Discipline"; Paul and Kimmelman, "Mendel in America"; Fitzgerald, *Business of Breeding*; Rosenberg, *No Other Gods*, ch. 13; Allen, "Reception of Mendelism." It bears mentioning that even among agriculturalists, Mendelism was differently promoted and received in different settings, at the level of institution and nation. For an overview of the international literature on Mendelism, see Charnley and Radick, "Intellectual Property."

6. For an account of de Vries's early life and career, see Schwartz, *In Pursuit*, ch. 4.

7. De Vries, *Die Mutationstheorie*. This was soon published in English: de Vries, *Mutation Theory*.

8. On the development of mutation theory, see Stamhuis, Meijer, and Zevenhuizen, "Hugo de Vries." See also Allen, "Hugo de Vries."

9. Stamhuis, Meijer, and Zevenhuizen, "Hugo de Vries"; Allen, "Hugo de Vries."

10. De Vries, *Mutation Theory*, 1:iix–ix.

11. Kingsland, "Battling Botanist." On experimentalism and American biology, see Pauly, *Controlling Life*; Benson, "Transformation of Natural History."

12. Kohler, *Lords of the Fly*. See also Allen, "Thomas Hunt Morgan"; Bowler, "Hugo De Vries." Campos gives an alternative account; see Campos, *Radium*, 155–62.

13. De Vries, *Plant-Breeding*, v. For background on de Vries's commitment to practical endeavors, see Theunissen, "Knowledge Is Power."

14. E.g., "Grow Larger Grain," 4; "How to Increase World's Grains," 4. On media hype and the mutation theory, see Endersby, "Mutant Utopias."

15. De Vries, *Mutation Theory*, 1:x.

16. Smith, *Garden of Invention*, ch. 13. On Burbank, see also Dreyer, *Gardener Touched with Genius*; Pandora, "Knowledge Held in Common."

17. Smith, *Garden of Invention*, 90–93.

18. Ibid., 133–56. For a discussion of Burbank in relation to the science of the time, see Palladino, "Wizards and Devotees."

19. Smith, *Garden of Invention*, 181–86.

20. From D. B. Weir report to the *Pacific Rural Press*, quoted in ibid., 119.

21. Harwood, *New Creations*, 344–45. See also Harwood, "Wonder-Worker."

22. De Vries, "Visit to Luther Burbank." On the debate between de Vries and Burbank, see Kingsland, "Battling Botanist."

23. De Vries, "Luther Burbank's Ideas," 678.

24. Ibid., 680.

25. On popular responses to x-rays and radium in the United States, see Lavine, *First Atomic Age*.

26. Lavine, *First Atomic Age*; Kevles, *Naked to the Bone*, chs. 1 and 2.

27. Campos, *Radium*; Campos, "Birth of Living Radium."

28. Lavine, *First Atomic Age*; see also Caufield, *Multiple Exposures*, ch. 3.

29. Kingsland, "Battling Botanist"; Campos, *Radium*, ch 3.

30. Campos, *Radium*, 121–25.

31. Gager, *Rays of Radium*, 256. See also Campos, *Radium*, 131–53.

32. For one contemporary evaluation, see Pollock, "Physiological Variations."

33. Brewster, "Plants and Animals to Order," 9658. This sensationalized news report reflected contemporary trends in science reporting; on science journalism in the early twentieth-century United States, see LaFollette, *Making Science Our Own*.

34. Coulter, *Fundamentals*, 50, 52. See also Shull, "Mutation Theory."

Chapter Two

1. Fitzgerald, *Business of Breeding*, 30–42, 56–74.

2. For an explanation of mutation dating to this period, see Morgan, *Physical Basis of Heredity*, ch. 20.

3. Sinnott and Dunn, *Principles of Genetics*, 307–8, 369. For a similar assessment, see Babcock and Clausen, *Genetics in Relation to Agriculture*, 476. On the importance of studies of mutation within the history of biology, see Campos and von Schwerin, "Making Mutations."

4. For an explanation of linkage from the early twentieth century, see Morgan, *Physical Basis of Heredity*, chs. 6–8.

5. On the development of chromosome mapping, see Kohler, *Lords of the Fly*, ch. 3.

6. See contributions to Rheinberger and Gaudillière, *Classical Genetic Research*, esp. those by Falk, Gannett and Griesemer, and Kass and Bonneuil.

7. Bridges, "Linkage Variation in Drosophila."

8. Plough, "Effect of Temperature on Crossingover"; Detlefsen and Clemente, "Genetic Variation in Linkage Values"; Detlefsen and Roberts, "Studies on Crossing Over."

9. The history of the GE research laboratory is discussed in detail in chapter 5. On the impact of the Coolidge tube, see Kevles, *Naked to the Bone*, 63–64; Reich, *American Industrial Research*, 89–90; Wise, *Willis R. Whitney*, 178–79.

10. Hook, "James Watt Mavor," 280. See also Wise, *Willis R. Whitney*, 254–55.

11. For a full account of this research, see Hook, "James Watt Mavor." See also Mavor's publications, e.g., Mavor, "Elimination of the X-Chromosome"; Mavor, "Production of Non-Disjunction" (1922); Mavor, "Production of Non-Disjunction" (1924); Mavor, "Effect of X Rays"; Mavor and Svenson, "X-Rays and Crossingover."

12. Mavor, "Attack on the Gene."

13. Woolley, "Electricity," 46–47.

14. There is evidence to suggest that other scientists actively disputed or disparaged these results. For example, the responses to Mavor's work made Albert Blakeslee hesitant to claim radiation-induced mutation in *Datura*. See Blakeslee to Gager, 14 January 1923, Blakeslee Papers, Folder "Gager, Stuart #4."

15. Stadler to Mavor, 21 August 1923, UM Agronomy, Box 5, Folder "August 1923, I–Q."

16. Rhoades, "Lewis John Stadler."

17. Poehlman, *History of Field Crops*, 63.

18. On the early history of maize genetics, see Rhoades, "Early Years"; Coe, "Origins of Maize Genetics"; Coe, "Birth of Maize Genetics." See also Comfort, *Tangled Field*.

19. On the arguments made for using different organisms in agricultural genetics, see Kimmelman, "Organisms and Interests."

20. Stadler, "Variability of Crossing Over," 1. See also Stadler, "Intensity of Linkage."

21. For a detailed account of hybrid corn, see Fitzgerald, *Business of Breeding*.

22. Stadler to Kinney, 7 March 1923, UM Agronomy, Box 4, Folder "March 1923, D–L."

23. Stadler to Emerson, 19 December 1923, Cornell DPB, Box 6, Folder "Stadler, LJ/Columbia, MO"; Stadler to Ball, 3 June 1924, UM Agronomy, Box 5, Folder "June 1924, A–H."

24. Stadler to Robertson, 11 June 1924, UM Agronomy Box 5, Folder "June 1924, I–Z"; Stadler to Schreiner, 25 June 1924, UM Agronomy, Box 5, Folder "June 1924, I–Z."

25. Rédei, "Portrait of Lewis John Stadler," 10.

26. Stadler, "Variation in the Intensity of Linkage in Maize," Cornell DPB, Box 6, Folder "Stadler, LJ/Columbia, MO."

27. Stadler to Henke, 21 March 1925, UM Agronomy, Box 6, Folder "June 1925, H–O."

28. University of Missouri Agricultural Experiment Station, "Solving Farm Problems" (1926), 38.

29. Here and elsewhere, for research in the early period of x-ray-induced mutagenesis, I do not characterize the extent of treatment in terms of the dosage of radiation as there was no standard measure at the time.

30. Stadler, "Genetic Effects of X-Rays in Maize," 72.

31. Stadler to Ball, 2 May 1925, UM Agronomy, Box 6, Folder "May 1925."

32. Stadler, "Genetic Effects of X-Rays in Maize."

33. Richey to Stadler, 24 September 1925, NARA RG 54, 66/31/136, Folder "L. J. Stadler 1918–1930 Sorted." Reference to the "apparent mutations" of the summer of 1925 also appears in Emerson to Stadler, 3 April 1926, Cornell DPB, Box 18, Folder "Stadler, LJ/Columbia, MO."

34. University of Missouri Agricultural Experiment Station, "Solving Farm Problems" (1927), 69–72.

35. Each seed of barley, carrying the dormant embryo of a future barley plant, contains within it cells that are already differentiated and which will grow into separate tillers of an adult plant; each tiller will in turn produce a head of self-fertilizing flowers. Stadler anticipated that if a genetic mutation were to appear in one cell of an irradiated barley seed, it would appear only in the tiller to which that cell gave rise and segregate only in the progeny of that particular tiller. See Stadler, "Mutations in Barley."

36. Stadler, "Mutations in Barley"; see also Stadler, "Some Genetic Effects of X-Rays in Plants."

37. Stadler to Richey, 8 June 1927, NARA RG 54, 66/31/136, Folder "L. J. Stadler 1918–1930 Sorted."

38. E.g., Emerson to Anderson, 1 June 1926, Cornell DPB, Box 8, Folder "E. G. Anderson/ Ann Arbor, Mich."

39. Goodspeed, "William Albert Setchell." See also correspondence related to the tobacco collection in Setchell Papers, Series 1, Folders "Goodspeed, T. H., 1909–1926" and "East, E. M. 1909–1918."

40. Jenkins, "Roy Elwood Clausen." On Goodspeed, see Baker, Foster, and Proskauer, "Thomas Harper Goodspeed."

41. Goodspeed, "X-Ray in Evolution," 226.

42. Goodspeed, "Effects of X-Rays and Radium," 243.

43. Gibson, Stewart, and Rollefson, "Axel Ragnar Olson." See also Olson, "X-Rays on Chemical Reactions"; Olson, Dershem, and Storch, "X-Ray Absorption Coefficients"; Beckwith, Olson, and Rose, "X-Ray Upon Bacteriophage."

44. Goodspeed, "X-Ray in Evolution," 227.

45. Goodspeed and Olson, "Production of Variation," 66. There are no significant records related to x-ray research in the Goodspeed Papers. Smocovitis reports that the Soviet scientist Michael Navashin arrived at Berkeley in 1927 to study *Crepis* evolution with Ernest Babcock, bringing with him x-ray-induced mutants; the relationship of Navashin's x-ray work to that of Goodspeed (or other work in x-ray-induced mutation) is not specified. See Smocovitis, "Plant *Drosophila*," 319–20.

46. Goodspeed, "Cytological and Other Features"; see also Goodspeed, "X-Ray in Evolution."

47. Goodspeed and Olson, "Production of Variation," 67, 68.

Chapter Three

1. Muller, "Artificial Transmutation," 84.

2. On efforts to induce mutation, see Morgan, Bridges, and Sturtevant, "Genetics of Drosophila," 26–27. On the challenges to "proving" induced mutation in these experiments, see Carlson, *Genes*, ch. 10; Luis Campos offers an alternative account in *Radium*, chs. 4 and 5.

3. Descriptions of Muller's early contributions to *Drosophila* genetics, as seen from the perspective of *Drosophila* geneticists, can be found in Dunn, *Short History*, 139–74; Sturtevant, *History of Genetics*, 45–57, 62–66. See also Carlson, *Genes*.

4. Muller, "Applications and Prospects," Muller Papers, Series II, Box 1, Folder "Muller Mss Writings 1916." Emphasis in the original. The original document is edited; lines deleted in the archival copy are not included here.

5. Muller and Altenburg continued their collaboration in subsequent summers. See Carlson, *Genes*, 109–13; Schwartz, *In Pursuit*, 222–24.

6. For a detailed account of Muller's path to the x-ray experiments, see Campos, *Radium*, ch. 5.

7. The special stock was known as ClB; the X chromosome of the flies carried a crossover suppressor, a recessive lethal, and a dominant mutation for "bar" eyes. This genetic combination enabled Muller to detect new recessive lethals after a precise series of crosses by checking vials for the complete absence of males. For a description of ClB and an explanation of its use, see Carlson, *Genes*, 109–19, and 120–50.

8. For descriptions of Muller's initial research on x-ray-induced mutation, see Carlson, *Genes*, 135–50; Campos, *Radium*, ch 5.

9. Muller, "Artificial Transmutation," 85.

10. For a nuanced account of why Muller's experiments in induced mutation proved more convincing or enduring than prior work, see Campos, *Radium*, ch. 5.

11. Muller, "Artificial Transmutation," 84.

12. Ibid., 84, 87. See also Muller, "Method of Evolution," 505.

13. Gager and Blakeslee's jimson weed experiments are discussed further in chapter 6.

14. Gager and Blakeslee, "Chromosome and Gene Mutations."

15. "Reports of the Sessions of Sections and Societies," 125.

16. "General Reports of the Second Nashville Meeting," 82.

17. Ibid., 84.

18. Curtis, "Old Problems," 149.

19. "Feathers on Birds," 6.

20. Stadler to Coffman, 4 November 1927, NARA RG 54, 66/31/136, Folder "L. J. Stadler 1918–1930 Sorted."

21. Stadler, "Genetic Effects of X-Rays in Maize," 74.

22. Stadler, "Some Genetic Effects of X-Rays in Plants," 8.

23. Stadler to Sax, 17 December 1931, Stadler Papers, Folder 5.

24. Muller, "Artificial Transmutation," 83, 85.

25. Goodspeed and Olson, "X-Rayed Sex Cells," 46.

26. From opening note, "Statement by the Discoverer," in Thone, "Science Prize Winner." Muller appears to have participated in at least one effort to extend his research to a crop plant by studying pecan nuts with a USDA breeder; see Traub and Muller, "X-Ray Dosage."

27. Muller, [untitled manuscript, ca. 1927], Muller Papers, Series II, Box 1, Folder "Muller Mss Writings, ca. 1927."

28. Slosson to Stadler, 31 May 1928, and Stadler to Slosson, 7 June 1928, Stadler Papers, Folder 1.

29. Stadler, "Some Genetic Effects of X-Rays in Plants," 18.

30. Ibid., 18–19.

31. "Tells Effect of Radium," 3.

32. "New Plant Varieties, Improved Animal Breeds," 6.

33. This stock article was copyrighted, and probably first circulated, in 1928; however, the version I discovered appeared in 1930. See Thone, "New Plants and Animals Developed."

34. "Scientists Grow Peculiar Tobacco Plants."

35. Muller, [untitled manuscript, ca. 1927], Muller Papers, Series II, Box 1, Folder "Muller Mss Writings, ca. 1927"; "Crowds 150 Years of Fly Evolution," 38.

36. "X-Rays Imperil Heredity," 4. See also Davis, "Magic Wand of X-Ray."

37. "X-Ray Speeds Up Nature," 5.

38. Thone, "X-Rays Speed Up Evolution," 243. For similar descriptions, see "Speeds Breeding Types"; "Plant Evolution Can Be Effected"; "X-Rays New Aid to Nurserymen."

39. "Species Improved." See also "Says X-Rayed Eggs Hatch Mostly Hens"; "Dieffenbach, Radiotherapy Specialist Dies."

40. Brosemer quoted in "X-Ray to Breed Super Animals," 6.

41. Gray, "New Life," 3.

42. Early, "Giant Flowers," 6. A similar extrapolation can be found in Gleason, "Sensational Study of Heredity," 17.

43. E.g., "Men and Things"; Frank, "No-Man's Land," 638, 642.

44. Serviss, "Transforming the World of Plants," 64. Emphasis in original.

45. Early, "Giant Flowers," 6. Similar assessments appeared elsewhere: Curtis, "Old Problems," 147; Thone, "X-Rays Speed Up Evolution."

46. Quotation in Early, "Giant Flowers," 6.

47. Gray, "New Life"; "X-Ray to Breed Super Animals."

48. "Secret of Life Sought," 567.

49. Muller, [untitled manuscript, ca. 1927], Muller Papers, Series II, Box 1, Folder "Muller Mss Writings, ca. 1927."

50. Crew quoted in Stokley, "Sir Oliver Lodge Expounds," 186.

51. The classic account of mass production is Hounshell, *American System*; a useful revisionist history is Scranton, *Endless Novelty*. See also Hughes, *American Genesis*, chs. 5 and 6.

52. Hounshell, introduction to *American System*, esp. 10–12.

53. Fitzgerald, introduction to *Every Farm a Factory*.

54. Danbom, *Born in the Country*, chs. 9 and 10.

55. On the appeal of efficiency in progressive-era America, see Alexander, *Mantra of Efficiency*. See also Haber, *Efficiency and Uplift*.

56. "X-Ray Tests Expected to Effect New Plants," 6; Thone, "X-Rays Speed Up Evolution," 244; Gray, "New Life," 3.

Chapter Four

1. Committee on Effects of Radiation upon Living Organisms (hereafter, CERLO), "Cumulative Report 1928–1934," March 1935, UT NRC.

2. Crocker to Kellogg, 13 June 1928, CERLO Files, Folder "Beginning of Program."

3. Curtis to Quin, 11 September 1928, CERLO Files, Folder "Request for Support Foundations."

4. On the role of foundations in science funding in this period, see Kohler, *Partners in Science*.

5. Samms to Curtis, 21 June 1928, CERLO Files, Folder "Requests for Support General."

6. CERLO, "Annual Report," April 1931, CERLO Files, Folder "Annual Report 1931." See the list of donors on pages 6–8.

7. Whitney to Curtis, 8 August 1928, CERLO Files, Folder "Request for Support General."

8. CERLO, "Annual Report," April 1931, CERLO Files, Folder "Annual Report 1931."

9. Wappler Electric Company to Curtis, 27 December 1928, CERLO Files, Folder "Requests for Support General."

10. For details of grantees and their projects, see various "Annual Reports" folders in the CERLO Files. A later publication that offers details on much of this research is Duggar, *Biological Effects of Radiation*.

11. Carlson, *Genes*, 19, 151–59.

12. Rhoades, "Lewis John Stadler." See also Roman, "Diamond in a Desert."

13. Stadler to Slosson, 7 June 1928, Stadler Papers, Folder 1; Stadler to Richey, 2 September 1929, NARA RG 54, 66/31/136, "L. J. Stadler 1918–1930 Sorted."

14. See mention in Bishop, "Mutations in Apples," 99. The study is also mentioned in Stadler, "Some Genetic Effects of X-Rays in Plants," 19.

15. Stadler to Clark, 16 May 1932, NARA RG 54, 66/31/136, Folder "L. J. Stadler 1932–1933."

16. California Agricultural Experiment Station, *Report* (1929), 11.

17. California Agricultural Experiment Station, *Report* (1930), 82.

18. On the history of the Berkeley Division of Genetics, see Kimmelman, "A Progressive Era Discipline," ch. 6.

19. California Agricultural Experiment Station, *Report* (1930), 82.

20. Published accounts of this research include Goodspeed, "Cytological and Other Features"; Goodspeed, "Effects of X-Rays and Radium"; Goodspeed, "Inheritance in *Nicotiana Tabacum*"; Goodspeed, "Meiotic Phenomena"; Goodspeed, "Triploid and Tetraploid Individuals."

21. CERLO, "Popular Report of Research Supported during year 1929–1930," n.d., UT NRC.

22. E.g., Goodspeed to Clayton, 25 June 1936, Goodspeed Papers, Box 1, Folder "Clayton, Edward Easton."

23. CERLO, "Cumulative Report 1928–1934," March 1935, UT NRC, 26.

24. McKay and Goodspeed, "Effects of X-Radiation," 644.

25. Smocovitis, "Plant *Drosophila*."

26. CERLO, "Popular Report of Research Supported during year 1929–1930," n.d., UT NRC, 7.

27. Horlacher and Killough, "Radiation-Induced Variation."

28. Horlacher and Killough, "Progressive Mutations Induced," 535–36.

29. Horlacher and Killough, "Radiation-Induced Variation," 256.

30. Horlacher and Killough, "Progressive Mutations Induced," 535.

31. Horlacher and Killough, "Radiation-Induced Variation," 256, 260.

32. Ware, "Plant Breeding," 742–43.

33. "Unusual Possibilities in Breeding," 183.

34. Martin, "Sorghum Improvement," 538.

35. Garner, "Superior Germ Plasm," 804, 828.

36. Jones, "Improvement in Rice," 441.

37. Unfortunately, research by private companies into the potential of induced-mutation breeding remains difficult to trace via archival sources. My examples come from newspaper, magazine, and journal articles.

38. Foster, "Artificially Induced Mutations," 23.

39. "Burpee Introduces," 29; "First X-Ray Flowers," 22. See also *Burpee Seeds* [catalog] (1942).

40. "Science: New Flowers by X-Rays."

41. Taylor, "Flowers Are Remodeled," 55.

42. GE's innovation activities are explored further in chapter 5. Background on David Burpee and W. Atlee Burpee & Co. is found in chapter 10.

Chapter Five

1. Among industrial firms, I am aware only of GE as having explored x-rays as a technology for plant breeding; however, other firms similarly saw the creation of technologies that would address farm needs as a source of growth. See, e.g., Boersma, "Organization of Industrial Re-

search." For a contemporary popular account of the GE induced-mutation research program, see "Magic-Ray Farmers."

2. On the history of the GE research laboratory in this period, see Reich, *American Industrial Research*; Wise, *Willis R. Whitney*. See also Birr, *Pioneering in Industrial Research*.

3. Wise, *Willis R. Whitney*, 131.

4. One standard account of this transition is given in Hughes, *American Genesis*. The literature on the history of American industrial research laboratories is extensive. For a review of works prior to 1990, see Smith, "Scientific Tradition in American Industrial Research." For more recent treatments, see Hounshell, "Evolution of Industrial Research"; Boersma, "Structural Ways."

5. Rasmussen, "Forgotten Promise"; Rasmussen, "Plant Hormones."

6. On the history of chemurgy, see Pursell, "Farm Chemurgic Council"; Beeman, "'Chemivisions'"; Finlay, "Old Efforts at New Uses."

7. Hawkins, "Electricity and Animation." On patterns of electricity development and use in America, see Nye, *Electrifying America*.

8. Wise, *Willis R. Whitney*, 178–79; Birr, *Pioneering in Industrial Research*, 152–53.

9. E.g., Davey, "Some Interesting Applications"; Davey, "Application of the Coolidge Tube"; Davey, "Effects of X-Rays."

10. Wise, *Willis R. Whitney*, 254–55.

11. Moore's patents at General Electric leading up to 1932 included a terminal for metal-sheathed wire (1913), a resistance element (1914), and an x-ray apparatus (1922). On Moore, see also Birr, *Pioneering in Industrial Research*, 60.

12. Nye, "Oral History," 3. For more on Haskins, see Dadourian, *Bio-Bibliography*.

13. "Magic Ray Farmers"; "Cathode Ray Yields Chemical Secrets."

14. Haskins and Moore, "X-Ray and Cathode Ray Tubes in the Service of Biology," 330.

15. "Symposium of Some Activities," 40.

16. Ibid., 39.

17. Haskins quoted in Gray, "New Life," 13. See also Haskins, "X-Ray and Cathode-Ray Tubes in Biological Service," 471.

18. E.g., Swingle, "New Citrous Fruits."

19. For publications derived from these and other studies, see, e.g., Moore and Haskins, "X-Ray Induced Modifications"; Haskins and Moore, "Inhibition of Growth"; Haskins and Moore, "Growth Modifications."

20. On similar activities conducted by RCA to support and extend the use of its electron microscope, see Rasmussen, "Making a Machine Instrumental."

21. Haskins quoted in Gray, "New Life," 3.

22. Haskins, "X-Ray and Cathode-Ray Tubes in Biological Service," 469.

23. E.g., Haskins and Moore, "Inhibition of Growth." Their published papers generally went above and beyond the details presented by their peers in biological research.

24. "X-Rays Are Found to Alter Heredity," 19.

25. Haskins quoted in Gray, "New Life," 3. With his vision of physics-based biology, Haskins was in step with a pattern of research increasingly evident throughout the discipline of biology and especially genetics. Physicists as well as ideas and tools drawn from the physical sciences came to play an important, though sometimes overstated, role in the unfolding of genetics research in the middle decades of the twentieth century. For a synthesis of the relevant literature, see Morange, *Molecular Biology*, ch. 7. Important contributions include Kohler, "Management of Science"; Keller, "Emergence of Molecular Biology"; Zallen, "Rockefeller Foundation"; Kay, *Molecular Vision*.

26. "Science: GE's Lily." See also Goldsmith, "X-Ray Produces Super-Lily"; "Better Regal Lily."

27. On plant patenting, see Bugos and Kevles, "Plants as Intellectual Property"; Fowler, "Plant Patent Act."

28. Pottage and Sherman, "Organisms and Manufactures."

29. Chester N. Moore, Regal Lily, US Plant Patent 165, filed 16 October 1934, and issued 18 February 1936.

30. Nye, "Oral History," 4. On the reconfigurations of the laboratory's research in this period, see Wise, *Willis R. Whitney*, 301–4 and charts on 246–47.

31. Kingsbury, *Hybrid*, ch. 13.

32. Andersen, "Sanilac Story."

33. CERLO, "Cumulative Report 1928–1934," March 1935, UT NRC, 25, 27 (quotation).

34. Kaempffert, "Science in the News."

35. Bowler, *Evolution*, 307–18.

36. Carlson, *Genes*, 166–69.

37. Haskins quoted in Gray, "New Life," 13.

38. Stadler to Slosson, 7 June 1928, Stadler Papers, Folder 1.

Part Two

1. "Chemical-Created Flower," 6; see also "Chemical Born Flower."

2. Burpee and Taylor, "So We Shocked Mother Nature," 15–17.

3. Thomas P. Hughes characterized this as hunt and try, presumably to avoid the connotation that there was any error involved in such a deliberate mode of testing; see Hughes, *American Genesis*, esp. ch. 1. A similar approach characterized certain areas of pharmaceutical innovation in which the systematic testing of an endless series of compounds against various disease agents was one route to new antibiotics and other chemotherapies. See, e.g., Hüntelmann, "Seriality and Standardization"; Lesch, *First Miracle Drugs*; Slater, "Malaria Chemotherapy."

4. The best overview of twentieth-century amateur technologists is Haring, *Ham Radio's Technical Culture*, ch. 1. The literature on amateurs is discussed further in chapter 9.

5. On tinkering with scientific instruments, see, e.g., Nutch, "Gadgets, Gizmos, and Instruments"; Schaffer, "Easily Cracked." Examples of experimenters tinkering with their experimental equipment are rife in the history of science and science studies literature; for an early theoretical take on tinkering, see Knorr, "Tinkering toward Success." On the importance of repair and maintenance in the history of technology, see Edgerton, *Shock of the Old*, ch. 4.

6. "Burpee Has Used Atomic Energy," 10.

Chapter Six

1. Belling, "Triploid and Tetraploid Plants," 463–64.

2. Davenport, "Annual Report of the Director" (1923), 98; Davenport, "Annual Report of the Director" (1926), 42.

3. For a detailed account of the development of ideas about polyploidy in plant evolution, see Smocovitis, "Botany and the Evolutionary Synthesis," esp. 107–17. For a fuller account of the use of *Oenothera* in biological research, see Endersby, *Guinea Pig's History*, ch. 5.

4. Biographical information on Gates is drawn from Roberts, "Reginald Ruggles Gates."

5. Gates, "Pollen Development in Hybrids," 110.

6. Ibid., 108–9.

7. Gates, *Mutation Factor in Evolution*, vi.

8. Gates, "Behavior of the Chromosomes."

9. Gates, "Polyploidy."

10. Lutz, "Preliminary Note." See also Richmond, "Women in Mutation Studies."

11. Anne Lutz's research findings, as noted in Gates, *Mutation Factor in Evolution*, 132–33.

12. Gates, "Mutations and Evolution (Continued)."

13. Ibid., 64–65.

14. For a review of this literature, see Gates, "Polyploidy," 164–66. See also Gates, *Mutation Factor in Evolution*, 197–203.

15. Gates, "Polyploidy," 166. See also Muller, "Why Polyploidy Is Rarer."

16. Winge's theory is described in Westergaard, "Øjvind Winge."

17. Blakeslee and Avery, "Adzuki Beans and Jimson Weeds." On Blakeslee's career, see Sinnott, "Albert Francis Blakeslee"; Kimmelman, "Mr. Blakeslee"; Campos, "Genetics without Genes"; Campos, *Radium*, ch. 4. The history of *Drosophila* research presents a similar story of the transition from teaching tool to model organism; see Kohler, *Lords of the Fly*, 33–37.

18. Sinnott, "Albert Francis Blakeslee," 4–9. On the early history of the station at Cold Spring Harbor, see Allen, "Eugenics Record Office." See also Witkowski, *Illuminating Life*.

19. Blakeslee to de Vries, 7 April 1933, Blakeslee Papers, Folder "de Vries, Hugo."

20. Blakeslee and Avery, "Mutations in the Jimson Weed."

21. On Belling, see Babcock, "John Belling"; Blakeslee, "John Belling."

22. Blakeslee, "John Belling," 83. On Belling's method, see Belling, "Iron-Acetocarmine Method."

23. Blakeslee, Belling, and Farnham, "Chromosomal Duplication."

24. Ibid., 389. They later elaborated on the results in Blakeslee and Belling, "Chromosomal Mutations."

25. Blakeslee and Belling, "Chromosomal Mutations."

26. For descriptions of the *Datura* research in relation to the history of cytogenetics, see Dunn, *Short History*, 158–59; Sturtevant, *History of Genetics*, 73.

27. For a further account, see Campos, "Genetics without Genes."

28. Blakeslee, "Types of Mutations," 263.

29. Gregory, "Genetics of Tetraploid Plants"; Keeble, "Giantism in *Primula Sinensis*."

30. Blakeslee did recognize a possible exception to this generalization in which a plant ended up with an additional homologous pair of chromosomes that presumably could pair and sort at meiosis in predictable ways. Blakeslee, "Types of Mutations," 265.

31. Blakeslee, "Variations in Datura," 31.

32. Belling, "Production of Triploid and Tetraploid Plants."

33. Blakeslee and Belling, "Chromosomal Mutations," 201.

34. On Davenport and eugenics, see Allen, "Eugenics Record Office"; Kevles, *In the Name of Eugenics*, ch. 3. See also contributions to Witkowski and Inglis, *Davenport's Dream*.

35. Their research and its publication are explained in detail in Campos, "Genetics without Genes"; see also Campos, *Radium*, 180–94.

36. Belling, "Production of Triploid and Tetraploid Plants," 463.

37. Davenport, "Annual Report of the Director" (1923), 98.

38. Davenport, "Annual Report of the Director" (1926), 42.

39. East, "Chromosome View of Heredity."

40. Sax, "Relation between Chromosome Number."

41. Randolph, "Some Effects of High Temperature," 223–24.

42. Dorsey, "Induced Polyploidy," 159.

43. Randolph, "Some Effects of High Temperature," 229.

44. A 1936 state-of-the-field review of polyploidy is Müntzing, "Evolutionary Significance of Autopolyploidy."

45. Perhaps the strongest evidence for Winge's hypothesis had come from Roy Clausen and Thomas Goodspeed, who had discovered a fertile tetraploid hybrid of a typically sterile pairing of tobacco species in the mid-1920s. See Clausen and Goodspeed, "Interspecific Hybridization."

46. Sax, "Experimental Production of Polyploidy"; Sax, "Effect of Variations in Temperature."

47. Kostoff, "Chromosome Alterations by Centrifuging," 302.

48. Imperial Bureau of Plant Genetics, *Experimental Production of Haploids and Polyploids*, 5.

Chapter Seven

1. Bates, "Polyploidy Induced," 316.

2. Blakeslee, "New Jimson Weeds," 85.

3. Ibid., 104.

4. Ibid., 107.

5. "Dearth of Datura Foreseen by All: Startling Facts here disclosed for the first time," memorandum, n.d., Blakeslee Papers, "Blakeslee, AF, Lectures, Papers, Etc: Miscellaneous Notes and Fragments."

6. Porter, "Prince's Poison."

7. On the history of experimentation with colchicine, see Goodman, "Plants, Cells, and Bodies." See also Eigsti and Dustin, *Colchicine in Agriculture*, 24–27.

8. For a more detailed account of how this knowledge traveled to Cold Spring Harbor, see Eigsti, "Cytological Study"; Eigsti and Dustin, *Colchicine in Agriculture*, 19–20. On the New York experiment station, see Nebel and Ruttle, "Cytological and Genetical Significance."

9. Eigsti to Blakeslee, 7 May 1935, and 15 April 1935, and Blakeslee to Eigsti, 15 August 1935, CIW Files, "O. J. Eigsti."

10. Eigsti, "Cytological Study." See also Eigsti and Dustin, *Colchicine in Agriculture*, 16–21.

11. Blakeslee and Avery, "Inducing Doubling of Chromosomes."

12. Ibid.

13. Because there were many researchers working with colchicine, the general cellular effects of which had been known for some time, disputes over priority for the "discovery" of colchicine-induced polyploidy did arise, especially from the Belgian researchers and Eigsti. See, e.g., Eigsti, Dustin, and Gay-Winn, "Action of Colchicine"; Havas, "Colchicine Chronology."

14. Woodburn, *20th Century Bioscience*, 60–62.

15. Wassermann, "Bernard Rudolf Nebel." Nebel wrote a short treatise in German on fruit cytology, which appeared in early 1937; see R.W., review of *Cytology of Fruits as Related to Plant Breeding*.

16. Nebel, "Cytological Observations."

17. Nebel and Ruttle, "Cytological and Genetical Significance."

18. Ibid., 9.

19. Eigsti, "Cytological Study," 63.

20. Nebel and Ruttle, "Colchicine and Its Place," 12.

21. Blakeslee and Avery, "Inducing Doubling of Chromosomes," 409.

22. Ibid., 408.

23. Ibid., 410. As mentioned, Blakeslee's use of the term "genetics engineer" never caught on; I discuss the late twentieth-century emergence of genetic engineering in the epilogue.

24. Nebel and Ruttle, "Cytological and Genetical Significance," 9.

25. Nebel and Ruttle, "Colchicine and Its Place," 15.

26. On science news reporting in the 1930s (with particular attention to chemistry), see LaFollette, "Taking Science to the Marketplace," esp. 261–62. See also Tobey, *American Ideology*; LaFollette, *Making Science Our Own*; LaFollette, *Science on the Air*.

27. "Control over Fundamental Life Processes," 20.

28. David, "New Elixir Found," 17.

29. "Old Gout Remedy," 9.

30. "Science Opens the Way," 7. The story circulated in many papers as it appeared in the syndicated Sunday supplement *American Weekly*. See also Editorial, "Colchicine and Double Diploids."

31. Blakeslee to Jones, 3 February 1938, Blakeslee Papers, Folder "Jones, DF 4."

32. E.g., "Evolution to Order," transcript of Science Service radio broadcast, 24 March 1938, Blakeslee Papers, Folder "Lecture, Papers, Etc.: Adventures in Science."

33. Blakeslee to Richey, 30 October 1937, CIW Files, Folder "USDA Bureau of Plant Industry."

34. "Evolution to Order," transcript of Science Service radio broadcast, 24 March 1938, Blakeslee Papers, Folder "Lecture, Papers, Etc.: Adventures in Science."

35. Nebel quoted in "Ancient Gout Treatment," 9.

36. Dietz, "Drug Creates Bigger Fruits," 45.

37. Blakeslee, "Annual Report of the Director" (1939), 181.

38. Blakeslee to Stout, 3 September 1937, CIW Files, "Blakeslee, Albert 1937."

39. Blakeslee to Richey, 3 September 1937, CIW Files, "USDA Bureau of Plant Industry."

40. Richey to Blakeslee, 2 October 1937, CIW Files, "USDA Bureau of Plant Industry."

41. Blakeslee to Sievers, 2 July 1938, CIW Files, "USDA Bureau of Plant Industry."

42. Sievers to Blakeslee, 19 July 1938, CIW Files, "USDA Bureau of Plant Industry."

43. Blakeslee to Moe, 14 January 1938, CIW Files, "Blakeslee, Albert Jan–June 1938." See also Blakeslee to Moore, 15 November 1937, CIW Files, "Blakeslee, Albert 1937."

44. Blakeslee, "Department of Genetics, Chromosome Investigations," 39.

45. Blakeslee to Richey, 3 September 1937, CIW Files, "USDA Bureau of Plant Industry."

46. Blakeslee's more entrepreneurial efforts fit well with the picture of him drawn by Barbara Kimmelman in "Mr. Blakeslee."

47. Nebel and Ruttle, "Cytological and Genetical Significance," 9.

48. Nebel, "Inducing Changes," 10.

49. Nebel and Ruttle, "Cytological and Genetical Significance," frontispiece.

50. Nebel, "Inducing Changes," 10.

51. Ibid., 15.

52. Way, *History of Pomology*.

53. Nebel and Ruttle, "Colchicine and Its Place," 8–10.

54. Ruttle and Nebel, "Extend Use of Colchicine," 10.

55. Nebel and Ruttle, "Colchicine and Its Place," 16–17.

56. Ruttle, "Polyploid Essential Oil Plants."

57. Nebel and Ruttle, "Colchicine and Its Place," 12.

58. Ruttle and Nebel, "Extend Use of Colchicine," 10.

Chapter Eight

1. Wellensiek, "Newest Fad."

2. E.g., Havas, "Colchicine Chronology," 115.

3. Eigsti, *Colchicine Bibliography*.

4. McDonough, "Agriculture Dept.," SM11; Bliven, "Remaking the World," 657.

5. Sears to East, n.d. [in response to a letter of 1 April 1937], Sears Papers, Folder 155.

6. Sears, "Amphidiploids in the *Triticinae*."

7. On wheat-rye hybrids, see Larter, "Historical Development."

8. "Use of a Chemical in Plant Breeding."

9. Dermen, "Colchicine Polyploidy and Technique," 600.

10. Sears to East, 1 April 1937, Sears Papers, Folder 155.

11. Baker, "Induced Polyploid," 187.

12. Dermen, "Colchicine Polyploidy and Technique"; Eigsti, *Colchicine Bibliography*; Randolph, "Evaluation of Induced Polyploidy."

13. Wellensiek, "Newest Fad."

14. Bates, "Polyploidy Induced," 315.

15. McManus, "Agriculture's Giant Powder," 13.

16. Taylor, "Nature Gets the Speed Up," 102.

17. McDonough, "Agriculture Dept.," SM11.

18. Teale, "Test-Tube Magic," 61, 227.

19. Oliver, "Horizons in Test Tubes," E4.

20. Laurence, "Science in the News," D4.

21. On the growth of the American chemical industry in the twentieth century, see Chandler, *Industrial Century*, chs. 3–6; Mowery and Rosenberg, *Paths of Innovation*, ch. 4. For a general history of the chemical industries, see Aftalion, *International Chemical Industry*.

22. For the origins of this slogan in DuPont advertising, see Meikle, *American Plastic*, 134; an overview of plastic types developed in the 1930s and the consumer products they made possible is given on 74–90.

23. For an excellent illustration of this, see Morrison, *Man in a Chemical World*.

24. Gardner, *American Agriculture*, 22–23.

25. Ibid., 24–25. See also Russell, *War and Nature*; Buhs, *Fire Ant Wars*.

26. Rasmussen, "Plant Hormones"; Anderson, "War on Weeds."

27. Harding, "Science and Agricultural Policy," 1098.

28. "Growing 'Em Bigger," 680.

29. Interest of Dow mentioned in a letter from H. E. How, editor of the *Journal of Industrial and Engineering Chemistry*; see How to Blakeslee, 13 September 1937, CIW Files, Folder "Blakeslee, Albert 1937."

30. "Growing 'Em Bigger."

31. "Plants Grown to Order," 58; de Lourdes, "Growth-Regulating Substances," 151; "Cambridge Laboratories" [advertisement].

32. Lesch, *First Miracle Drugs*.

33. "Wonder Plant-Drug," 2.

34. "'Shots in Arm,'" 10. See also J. W. B., "Lily Gets a Hypo."

35. "Plants Doped," 8. See also "Bigger and Better Berries"; "Tricking Dame Nature."

36. For a history of the rubber shortage, see Tuttle, "Birth of an Industry." See also Finlay, *Growing American Rubber.*

37. Warmke and Davidson, "Polyploidy Investigations" (1942); Warmke and Davidson, "Polyploidy Investigations" (1943). See also Warmke, "Experimental Polyploidy."

38. On the American home front in World War II, see Adams, *Best War Ever*; Jeffries, *Wartime America.*

39. Bentley, *Eating for Victory*, 117.

40. Sonnedecker, "Fuel for Fighters," 156.

41. "War Gardens," 16.

42. De Lourdes, "Growth-Regulating Substances," 152.

43. Prather, "Better Nutrition," 2–3.

44. "Plant Magicians," 94, 99.

45. Danbom, *Born in the Country*, 233–44.

46. There is a significant literature on the Rockefeller Foundation's agricultural programs and the Green Revolution. On the Mexico program that began in the 1940s, often considered the starting point of the Green Revolution, see Fitzgerald, "Exporting American Agriculture"; Cotter, *Troubled Harvest*, chs. 4–7; Harwood, "Peasant Friendly Plant Breeding." On the broader geopolitical context, see Perkins, *Geopolitics*; Cullather, *Hungry World.*

47. Cooley, "Apples as Big as Your Head!," 79.

Chapter Nine

1. "Control over Fundamental Life Processes," 20.

2. Taylor, "Nature Gets the Speed Up," 100.

3. Couch, "Botanist Upsets Heredity," 11.

4. I use the term "amateur" to describe those pursuing work in plant breeding and genetics who were not employed to do so professionally. This assemblage included expert amateurs (for example, members of flower societies or farmers knowledgeable about crop breeding) as well as novice amateurs.

5. On amateur radio and television operators, see Douglas, "Audio Outlaws"; Douglas, *Inventing American Broadcasting*, ch. 6; Takahashi, "Network of Tinkerers"; Haring, "Freer Men"; Haring, *Ham Radio's Technical Culture.* On computer tinkering, see Tinn, "From DIY Computers"; see also Kelly-Campbell and Aspray, *Computer*, ch. 10. For an account of early amateur rocketry in the United States, see Springer, "Early Experimental Programs." Other accounts of tinkering in the history of technology include Waksman, "California Noise"; Franz, *Tinkering.*

6. Blakeslee and Avery, "Inducing Doubling of Chromosomes," 404–5.

7. Thone, "Science Stunts," 235.

8. See, e.g., "X-Ray Produces Rare Gladiolus"; "Musician Speeds Up Tree Growth."

9. Blakeslee, form letter regarding colchicine, 6 December 1939, Blakeslee Papers, Folder, "Colchicine—Correspondence 1."

10. Eigsti and Tenney, *Report on Experiments*, 21.

11. "Reported from the Field of Science," 59. Other researchers similarly received an influx of colchicine-related requests from individuals eager to participate in the research. For example, Ernie Sears of the University of Missouri heard from many such persons after publishing an article on colchicine in 1939. See Sears Papers, Folder 116 (and elsewhere in the collection).

12. Blakeslee, form letter regarding colchicine, 6 December 1939, Blakeslee Papers, Folder, "Colchicine—Correspondence 1."

13. "Evolution to Order," transcript of Science Service radio broadcast, 24 March 1938, Blakeslee Papers, Folder "Lecture, Papers, Etc.: Adventures in Science."

14. For Blakeslee, surviving letters are primarily from the late 1940s and 1950s and in response to continued reporting on colchicine. I have assumed their authorship and content are similar to those of letters arriving in the 1930s.

15. King to Sears, 17 April 1939, Sears Papers, Folder 116.

16. Ratliff to "Gentlemen," 5 December 1945, Blakeslee Papers, Folder "Colchicine—Correspondence 2."

17. Abraham to Blakeslee, n.d. [rec'd 14 September 1950], Blakeslee Papers, "Colchicine—Correspondence 4."

18. Carson to Blakeslee, 20 April 1946, Blakeslee Papers, Folder "Colchicine—Correspondence 2."

19. Manzer to Avery and Blakeslee, 5 December 1945, Blakeslee Papers, Folder "Colchicine—Correspondence 2"; Swaney to Blakeslee, 18 January 1939, CIW Files, unfiled items "Swaney"; Blakeslee to Hartman, 14 November 1946, Blakeslee Papers, Folder "Colchicine—Correspondence 3."

20. E.g., Ehrhard to Avery and Blakeslee, 5 January 1947, Blakeslee Papers, Folder "Colchicine—Correspondence 3."

21. Manzer to Avery and Blakeslee, 5 December 1945, Blakeslee Papers, Folder "Colchicine—Correspondence 2"

22. MacDougall to Blakeslee and Avery, 1 December 1945, Folder "Colchicine—Correspondence 2."

23. Clark to "Sir," n.d. [ca. September 1953], CIW Files, Folder "Requests for Misc. Information 1943–1953."

24. Reps to Carnegie Institution, 9 August 1946, Blakeslee Papers, Folder "Colchicine—Correspondence 3."

25. McManus, "Agriculture's Giant Powder," 13.

26. Thone, "Amateur Plant Breeders Aid Science."

27. "Cambridge Laboratories" [advertisement].

28. Robert G. Cook, form letter on colchicine from *Journal of Heredity*, 1941, Blakeslee Papers, Folder "Blakeslee, Colchicine—Correspondence 1."

29. Logan, Putnam, and Cosper, *Science in the Garden*, 123–30, 241–42.

30. Richards to "Sir," 26 April 1947, CIW Files, Folder "Requests for Misc. Information, 1943–1953."

31. Houghton, "Experiments with Colchicine," 16.

32. Blanchard, "Colchicine in Gladiolus Breeding," 100.

33. For later examples see, e.g., Leach, "Some Notes"; Cresskill, "Supremes with Colchicine."

34. Bott, "Test Tube Garden."

35. E.g., Kennerly, "New Plants on Order." See also Haworth, *Plant Magic*; James, *Create New Flowers*.

36. Couch, "Botanist Upsets Heredity," 11.

37. "Chemical Is Offered."

38. Eigsti and Tenney, *Report on Experiments*, 7. See also the description in Ray, "Plant Magic Is World Hobby." Today this effort would be classed as "citizen science," an effort to engage nonscientists in research activities as a means of both educating participants and gathering additional data. On the long history of citizen-science-like activities, see contributions to a special issue on lay participation in scientific observation: *Science in Context* 24, no. 2 (2011), especially Jeremy Vetter's introduction.

39. Eigsti and Tenney, *Report on Experiments*, 6.

40. Ibid., 17, 19.

41. Ibid., 16.

42. Thone, "Amateur Plant Breeders Aid Science," 10.

43. Ray, "Plant Magic Is World Hobby," 175–76.

44. Editorial, "Colchicine and Double Diploids."

45. Editorial, "Colchicine a Dangerous Drug," 188.

46. "Colchicine Experimenters Warned," 382.

47. Morrison, "Facts about Colchicine," 297.

48. Rockwell, "'Round about the Garden," 38; Smith, "No Short-Cut Horticulture," 140.

49. McManus, "Agriculture's Giant Powder"; Gardner, "Giants in the Garden"; Bliven, "Remaking the World of Plants"; "Plant Magicians"; Stephenson, "New Plant World."

50. Blakeslee to Jones, 3 February 1938, Blakeslee Papers, Folder "DF Jones 4."

51. Cooley, "Apples as Big as Your Head," 79, images on 80 and 81.

52. Abraham to Blakeslee, n.d., Blakeslee Papers, Folder "Colchicine—Correspondence 4."

53. Turner to Carnegie Institution, 3 March 1949, Blakeslee Papers, Folder "Colchicine—Correspondence 4."

54. Douglas, *American Broadcasting*, ch. 6, esp. 206; Lindsay, "From the Shadows," 46–47.

55. Kennerly, "New Plants on Order," 236. Emphasis in original.

56. Bott, "Test Tube Garden."

Chapter Ten

1. Bates, "Polyploidy Induced," 316. See also Emsweller and Ruttle, "Induced Polyploidy in Floriculture."

2. A useful entry point into the literature on users in the history of technology is Oudshoorn and Pinch, *How Users Matter*, esp. the introduction. See also discussion in the introduction to this book.

3. Ken Kraft provides a lighthearted history of Burpee Seed in *Garden to Order*. A brief history is also available on the company's website in "Company History," Burpee.com, accessed 11 August 2014, http://www.burpee.com/gardening/content/company-history/history.html.

4. "Burpee for Burbank."

5. Taylor, "Flowers Are Remodeled," 52.

6. Teale, "Test-Tube Magic," 59.

7. Description of Floradale from *Burpee Seeds* [catalog] (1939).

8. Kraft, *Garden to Order*, 94–98.

9. Burpee and Taylor, "So We Shocked Mother Nature," 15.

10. Taylor, "Flowers Are Ornery Critters," undated MS, Taylor Papers, Box 4, Folder, "Flowers, They've Grown Some New Posies for the Ladies, SEP."

11. Taylor, "New Posies," 135.

12. Bugos and Kevles, "Plants as Intellectual Property."

13. Burpee's navigation of this competitive marketplace is described in Taylor, "Flowers Are Remodeled."

14. Kraft, *Garden to Order*, 73. According to the company history, David Burpee pioneered the use of first-generation hybrids in the flower-seed industry. See "Company History" in note 3. On intellectual property in plants, see Bugos and Kevles, "Plants as Intellectual Property."

15. Taylor, "Speeding Up Nature," 70.

16. Kraft, *Garden to Order*, 76–80.

17. Kingsbury, *Hybrid*, 180.

18. "Our men at Floradale" [Burpee employees], "Tetraploid Flowers," encl. in Burpee to Taylor, 12 December 1950, Taylor Papers, Box 4, Folder "Flowers, They've Grown Some New Posies for the Ladies, SEP."

19. The literature on the relationship between science and the consumer marketplace in the twentieth century is fragmented, but studies suggest the increasing use of science in advertising in the early decades of the twentieth century. One detailed historical study of how scientific claims were deployed in advertising in this period is Apple, *Vitamania*. On technology and modernity as represented in midcentury American advertising more generally, see Marchand, *Advertising the American Dream*. On marketing scientific and medical products, see Rentetzi, "Packaging Radium."

20. Waldron, "Turnips or Tulips," 140.

21. *Burpee Seeds* [catalog] (1942): 9.

22. Taylor, "New Posies," 137.

23. *Burpee Seeds* [catalog] (1949): 16.

24. On technological enthusiasm, see Hughes, *American Genesis*; Wright, *Possible Dreams*. On technological enthusiasm and amateur activities, see Douglas, "Audio Outlaws"; Waksman, "California Noise."

25. For a brief account of triticale research in relation to the history of cytogenetics, see Santesmases, "Cereals, Chromosomes and Colchicine." See also National Research Council, *Triticale*.

26. Kraft, *Garden to Order*, 94, 97.

27. Burpee to Denison, 2 September 1964, CIW Files, "Blakeslee, Albert 1940–1949."

28. A meandering account of Eigsti's career can be found in Woodburn, *20th Century Bioscience*.

29. Zorn, "Seedless Melons."

30. Severson, "Watermelons Get Small."

31. "Colchiploidy" was a phrase that the USDA breeder Haig Dermen especially liked to use, e.g., Dermen, "Colchiploidy in Grapes"; Dermen, "Colchiploidy and Histological Imbalance."

32. E.g., Riley, *Dahlias*, 149–52; Thomson, "New Kinds of Plants."

33. Callaway, *World of Magnolias*, 191–93; Clarke, *Marijuana Botany*, 61–62.

Part Three

1. The condition of this field (the "gamma field") and its operation in 1957 are described in detail in Lang, "Stroll in the Garden."

2. On the early history of the BNL, see Crease, *Making Physics*.

3. A general account of large technological systems is Hughes, "Evolution of Large Technological Systems."

4. On the development of electrical power systems, see Hughes, *Networks of Power*. See also Nye, *Consuming Power*.

Chapter Eleven

1. This was not the case in all countries. In Sweden, for example, a group of plant breeders carried out research into mutation breeding beginning in the 1930s and by the 1950s claimed

to have produced a number of improved types. See van Harten, *Mutation Breeding*, 54–56. The researchers whose work I discuss in part 3 were primarily interested in two forms of radiation: gamma rays emitted by radioisotopes and neutron radiation (here typically produced within a nuclear reactor). However, the surge in interest in these nuclear-research-derived forms of radiation prompted a renewed interest in other forms of radiation including the whole range of electromagnetic radiation (e.g., x-rays, ultraviolet).

2. A detailed account of the AEC, including its formation and early activities, can be found in its official histories: Hewlett and Anderson, *New World*; Hewlett and Duncan, *Atomic Shield*; Hewlett and Holl, *Atoms for Peace and War*. See also Mazuzan and Walker, *Controlling the Atom*.

3. Westwick characterizes the national laboratories as an "institutional system" along the lines of a technological system. See Westwick, *National Labs*, 7.

4. Creager, *Life Atomic*. For a short overview of AEC efforts to promote peaceful uses of atomic energy, see Boyer, *Bomb's Early Light*, ch. 24. See also references on radioisotope distribution in note 28.

5. On the production of the first atomic weapons, see Rhodes, *Atomic Bomb*. On the development of this atomic infrastructure, see also Hughes, *American Genesis*, 381–442; Westwick, *National Labs*, esp. ch. 1.

6. On the establishment of the AEC, see Hewlett and Anderson, *New World*, esp. 482–530. On the efforts of scientists in particular to prevent military control of nuclear science, see Smith, *Peril and a Hope*; Wang, *American Scientists*, ch. 1.

7. US Congress, Atomic Energy Act of 1946, Public Law 79–585.

8. For an overview of research supported in the early years of the AEC (including in biology and medicine), see Hewlett and Anderson, *New World*, 233–70.

9. Westwick, *National Labs*, 8–9.

10. Laurence, "Atomic Laboratory," 17.

11. Laurence, "Atomic Laboratory"; many of these projects and their gradual unfolding are described in Crease, *Making Physics*.

12. Curtis, "Leslie F. Nims." On the slow start to the biology program, see Crease, *Making Physics*, esp. 61–68.

13. Crease, *Making Physics*, 63; see also Nims, "Opportunities for Research."

14. BNL, *Conference on Biological Applications*.

15. Galinat, "Willard Ralph Singleton."

16. Singleton to Hollaender, 22 January 1948, Singleton Papers, Box 5.

17. Singleton to Shull, 22 January 1948, Singleton Papers, Box 5.

18. Ibid.; Nims to Singleton, 9 March 1948, Singleton Papers, Box 5.

19. Conger, "Arnold Hicks Sparrow."

20. On the experimental setup in the field's first year of use, see Sparrow, "Tradescantia Expt. Gamma Field," n.d. [1949/50], Sparrow Papers, Box 5, Binder "Gamma, 1949–1950."

21. There are many published descriptions of the gamma field. See, e.g., Shapiro, "Brookhaven Radiations Mutation Program"; Singleton, *Nuclear Radiation*, ch. 26; Sparrow and Singleton, "Radiocobalt as a Source of Gamma Rays." A more colorful account is Manchester, "'Atomic Crops.'"

22. Singleton, "Progress Report," 15 June 1949, Singleton Papers, Box 6; Nims to Haworth, 2 December 1949, AIP Brookhaven, Reel 9, Folder 10.

23. Singleton, "Progress Report," 15 June 1949, Singleton Papers, Box 6.

24. Sparrow, "Tolerance of Tradescantia . . ." n.d. [1949/50], Sparrow Papers, Box 5, Binder "Gamma, 1949–1950."

25. In addition to the experimental study of acute radiation effects in plants and animals, there was also ongoing research into the effects of long-term, low-dose exposure to radiation as experienced by workers in various fields and industries; see Walker, *Permissible Dose*. See also Hacker, *Dragon's Tail*.

26. For early experiments using radium on plants, see Campos, *Radium*, esp. ch. 3. For a characteristic collection of radiation studies from the 1920s and 1930s, see Duggar, *Biological Effects of Radiation*.

27. Creager, "Industrialization," 142–44.

28. Creager, *Life Atomic*. See also Lenoir and Hays, "Manhattan Project"; Creager, "Nuclear Energy"; Creager, "Phosphorus-32." On radioisotopes in ecology, see Hagen, *Entangled Bank*, ch. 6; Bocking, "Ecosystems, Ecologists"; Bocking, *Ecologists and Environmental Politics*, part 2. On the global distribution of radioisotopes (and the politics of this distribution), see Creager, "Tracing the Politics"; Creager, "Radioisotopes as Political Instruments"; Gaudillière, "Normal Pathways"; Krige, "Politics of Phosphorus-32"; Krige, "Atoms for Peace"; Santesmases, "Peace Propaganda."

29. The influence of radioisotopes on agricultural research (as opposed to biology, biomedicine, and ecology) is less well documented. One exception is food irradiation research; see, e.g., Buchanan, "Atomic Meal"; Zachmann, "Atoms for Peace"; Zachmann, "Risky Rays." An article that presents the AEC perspective on the use of radioisotopes in agricultural research is Oatsvall, "Atomic Agriculture."

30. Further sources on the influence of the AEC on life sciences research (beyond the case of radioisotope distribution) in the United States and abroad include Beatty, "Genetics in the Atomic Age"; Lindee, *Suffering Made Real*; Creager and Santesmases, "Radiobiology in the Atomic Age"; Rader, "Alexander Hollaender's Postwar Vision."

31. Sparrow and Singleton, "Radiocobalt," 29.

32. Stangby to Balber, "Tentative Handling of the 200 curies of Cobalt 60," 29 March 1951, Sparrow Papers, Box 5, Binder "Gamma Field: 1951." Emphasis in original.

33. This episode is recounted in O'Neil, "Yaphank's Happy Thinkers," 116.

34. Singleton, "Quarterly Progress Report," 2 October 1951, Singleton Papers, Box 6.

35. "Summary of Gamma Field Plants," 30 August 1951, Sparrow Papers, Box 5, Binder "Gamma Field: 1951."

36. Sparrow, [list of species irradiated, 1952 gamma field], n.d., Sparrow Papers, Box 6, Binder "1952 Gamma Field Experimental Record."

37. Singleton to Nims, 12 March 1948, Singleton Papers, Box 5.

38. For a summary of Stadler's perspective, see van Harten, *Mutation Breeding*, 50–51.

39. Singleton to Nims, 12 March 1948, and Singleton to Nims, 22 March 1948, Singleton Papers, Box 5.

40. "Scientist Converts Tall Field Corn," 1; BNL, *Annual Report, July 1, 1950*, 76.

41. Singleton, "Progress Report," 23 June 1950, Singleton Papers, Box 6.

42. On the publicity of peaceful uses of atomic energy more generally, see Weart, *Nuclear Fear*, ch. 8. See also Medhurst, "Atoms for Peace"; Titus, "Selling the Bomb."

43. Hecht, *Radiance of France*, 15–16.

44. On the intersection of science and politics in AEC support of the life sciences, see Beatty, "Scientific Collaboration"; Creager, *Life Atomic*; Creager, "Tracing the Politics"; Creager, "Radioisotopes as Political Instruments"; Krige, "Politics of Phosphorus-32"; Krige "Atoms for Peace"; Santesmases, "Peace Propaganda."

45. Considine, "Behind the Scenes." For an account that details (among other things) Singleton's involvement in promoting genetics and especially the contributions of Mendelian genetics to agricultural production, see Wolfe, "Cold War Context."

46. "New Knowledge Hot Corn."

47. "Atom Study Points to Food Plenty," 4.

48. US Atomic Energy Commission, *Some Applications*, 71–72.

49. Ibid., 93–95.

50. Ibid., 71.

51. Singleton, "Progress Report," 28 December 1951, Singleton Papers, Box 6.

Chapter Twelve

1. Singleton, "Progress Report," 23 June 1950, Singleton Papers, Box 6.

2. See letters of invitation, e.g., Curtis to Deering, 19 November 1952, AIP Brookhaven, Reel 9, Folder 10.

3. BNL, *Annual Report, July 1, 1953*, 44.

4. On the history of concerns about radiation and fallout, see Weart, *Nuclear Fear*; Winkler, *Life Under a Cloud*. A concise history of responses to known or perceived dangers from radiation is Walker, *Permissible Dose*. On radiation safety in the postwar period, see also Hacker, *Elements of Controversy*.

5. Shapiro, "Brookhaven Radiations Mutation Program," 143–45.

6. BNL, *Annual Report, July 1, 1952*, 88.

7. Shapiro, "Brookhaven Radiations Mutation Program," 148.

8. A list of cooperating institutions and species irradiated through 1954 can be found in Curtis to Tape, 4 February 1955, AIP Brookhaven, Reel 9, Folder 11.

9. See foldout in BNL, *Annual Report, July 1, 1954*. Quotation on page xi.

10. Ibid., 49.

11. E.g., Singleton, "Atomic Energy and Abundance," address delivered at the University of New Hampshire Chapter of Sigma Xi, 28 October 1954, Singleton Papers, Box 18.

12. Singleton to Mangelsdorf, 19 October 1953, Singleton Papers, Box 6.

13. Lundqvist, "Eighty Years."

14. Singleton to Meyers, 27 April 1953, Singleton Papers, Box 6.

15. Konzak, "Stem Rust Resistance."

16. Richter and Singleton, "Chronic Gamma Radiation."

17. Sites where induced-mutation research proved popular included the State College of Washington (later Washington State University), Iowa State University, North Carolina State University, and the University of Minnesota, among others.

18. O'Mara to Sears, 16 September [1953], Sears Papers, Folder 549.

19. O'Mara to Sears, 22 June 1954, Sears Papers, Folder 550.

20. See, e.g., a later discussion: Sears to O'Mara, 12 March 1962, Sears Papers, Folder 551.

21. O'Mara to Sears, 18 January 1954, Sears Papers, Folder 550.

22. Most objections were not openly about the support or influence of the AEC (though this may well have been a consideration for some), but rather about the rapid growth of interest in mutation breeding despite a lack of evidence for its usefulness and also the perceived lack of scientific or technical rigor found in many mutation breeding efforts. The objections of some breeders to the renewed interest in induced-mutation breeding, along with the re-

sponses of mutation breeders to these objections, is discussed in Leung, "Between Farming and Radioscience."

23. Dwight Eisenhower, address to the 470th Plenary Meeting of the United Nations General Assembly, 8 December 1953. Text online at https://www.iaea.org/about/history/atoms-for-peace -speech (accessed 26 April 2011).

24. John Krige identifies still other purposes of "Atoms for Peace," such as redirecting Soviet nuclear capabilities to the international program and circumscribing the development of nuclear capabilities in other countries to include only energy production and other nonmilitary activities. See Krige, "Atoms for Peace," 162–63. For a full account of "Atoms for Peace," see Hewlett and Holl, *Atoms for Peace and War*.

25. US Congress, Joint Committee on Atomic Energy, *Contribution of Atomic Energy*.

26. Ibid., 43–45.

27. Ibid., 55.

28. Blair, "Strauss Praises Atomic Crop Gain," 23. See a similar account by an AEC official in Libby, "Economic Potential," 5.

29. Although it is not entirely clear from the text, it seems likely that Hinshaw was referring to mutations in Japanese people exposed to the atomic detonations. US Congress, Joint Committee on Atomic Energy, *Contribution of Atomic Energy*, 52.

30. Walker, *Permissible Dose*, 10–18.

31. On scientific debates over the genetic effects of radiation, see Beatty, "Weighing the Risks"; Beatty, "Masking Disagreement Among Experts"; Hamblin, "Dispassionate and Objective Effort." On mutation in popular culture during the Cold War and after, see Masco, "Mutant Ecologies."

32. For a detailed account of the Atomic Bomb Casualty Commission, see Lindee, *Suffering Made Real*.

33. Beatty, "Genetics in the Atomic Age."

34. Moh, Nilan, and Elliott, "Unusual Association"; Smith, "Rare Dominant Chlorophyll Mutant"; "A-Bomb's Rays Produce Glowing Corn."

35. "Atom Alters Heredity," 188.

36. "Can A-Bomb Breed 4-Inch Monsters?" 35.

37. For a history of the test and the events that followed, see Hacker, *Elements of Controversy*, ch. 6.

38. Walker, *Permissible Dose*, 18–28.

39. Kopp, "Debate over Fallout Hazards." See also, on the continued debate among scientists, Beatty, "Weighing the Risks."

40. Edward F. Ryan, "Benefits of Nuclear Fission in Agriculture Are Cited," clipping from Washington, DC, newspaper, [1 or 2 April] 1954. From Singleton's clippings file; see Singleton Papers, Box 23.

41. "Radiation Used to 'Speed Up' Evolution," 5S. "Progress" should read "process."

Chapter Thirteen

1. See, e.g., Minutes, Advisory Committee for Biology and Medicine, Eleventh Meeting, 11 September 1948, OpenNet, accession no. NV0709065; Minutes, Advisory Committee for Biology and Medicine, Fifteenth Meeting, 8–9 April 1949, OpenNet, accession no. NV0708706.

2. According to Peter Westwick, in 1958 the AEC "supported up to one third of all [genetics] research in the United States"; see Westwick, *National Labs*, 2.

3. Report of the Tenth Meeting of the Advisory Committee for Biology and Medicine, 29 July 1948, OpenNet, accession no. NV0711607.

4. Fitzgerald, "Technology and Agriculture."

5. For an overview of research at Oak Ridge, see Johnson and Schaffer, *Oak Ridge.*

6. Johnson and Jackson, *City Behind a Fence,* 24–25.

7. Westwick, *National Labs,* 31–42.

8. Johnson and Schaffer, *Oak Ridge,* ch. 2.

9. *UT-AEC Agricultural Research Laboratory,* 3.

10. UTAES, *Sixty-Second Annual Report,* 157.

11. "UT-AEC Research Program," 3.

12. The system did not entirely work out as envisioned, and by the mid-1950s, AEC observers were concerned that there was not enough involvement of Oak Ridge National Laboratory staff. See, e.g., J. H. Jensen, H. A. Kornberg, and E. C. Stackman, "Report of a Special Committee to Study the UT-AEC Facility," 16 January 1957, OpenNet, accession no. NV0709121.

13. "UT-AEC Research Program," 10.

14. BNL, *Annual Report, July 1, 1954,* xxii–xxiii.

15. "UT-AEC Research Program," 10.

16. Westwick, *National Labs,* 10–23.

17. UTAES, *Sixty-Second Annual Report,* 158.

18. See the description and image of the field in UTAES, *Progress of Agricultural Research,* 36.

19. UTAES, *Sixty-Second Annual Report,* 161–72.

20. "Atom Ray Used on Seeds," 17.

21. "Atomic Oven Bakes Seeds."

22. Osborne to Stadler, 16 April 1953, Stadler Papers, Folder 141.

23. UTAES, *Sixty-Seventh Annual Report,* 21–22.

24. UTAES, *Sixty-Eighth Annual Report,* 71.

25. Osborne and Lunden, "Cooperative Plant and Seed Irradiation."

26. Reed and Osborne, "Soybean Research," 19.

27. UTAES, *Progress of Agricultural Research,* 37.

28. Johnson and Epps, "Radiation Tests," 4.

29. UTAES, *Progress of Agricultural Research,* 36–37.

30. *UT-AEC Agricultural Research Laboratory,* 38.

31. "Seeks Cross That Will Stay Crossed."

32. UTAES, *Sixty-Eighth Annual Report,* 70–71.

33. Minutes, Meeting of the Advisory Committee for Biology and Medicine, Fifty-First Meeting, 5–7 May 1955, OpenNet, accession no. NV0411745; Shoup to Roth, 19 May 1955, OpenNet, accession no. NV0706973.

34. Osborne and Lunden, "Cooperative Plant and Seed Irradiation," 199.

35. Ibid., 203–5.

36. Ibid., 208–9.

37. "Planters Now Can Predict," 2.

38. Osborne to Garrison, 23 April 1957, CU Horticulture, Series 88, Box 24, Folder 5.

39. Osborne and Lunden, "Cooperative Plant and Seed Irradiation," 208.

40. Osborne, "Regional and National Programs," 41–42.

41. Osborne, "Radiation and Plant Breeding," 8.

42. Osborne, "Atomic Tools," 9.

43. Teas, "Station Installs Cobalt Irradiator," 4; Teas, "Use of Cobalt-60."

44. Teas, "Station Installs Cobalt Irradiator," 4.

45. Romani et al., "Cobalt-60 Gamma-Ray Irradiator," 3.

46. Ibid., 2, 4.

Chapter Fourteen

1. Burpee to Taylor, 20 December 1950, Taylor Papers, Box 4, Folder "Flowers, They've Grown Some New Posies for the Ladies, SEP."

2. Brownell, *Seeds of Tomorrow*. See also "Garden Time," 1.

3. On early incorporation of atomic themes into popular culture, see Boyer, *Bomb's Early Light*, ch. 1. On commercial use of atomic imagery (in this case uranium-themed products), see Amundson, "Uranium on the Cranium."

4. See fig. 25; see also "Atomic-Energized Seeds and Plants" [advertisement].

5. On the history of atomic gardening in Great Britain, see Johnson, "Safeguarding the Atom"; or the 1960s guide, Howorth, *Atomic Gardening*.

6. Zeman and Amundson, introduction. On growing opposition to nuclear development, see Egan, *Barry Commoner*; Moore, *Disrupting Science*.

7. See discussion of tinkering in chapter 9.

8. "Speas Story."

9. As reported in "More Food with Atom-Blasted Seeds," 7.

10. "Speas Story."

11. "More Food with Atom-Blasted Seeds."

12. E.g., "Atomic Rays Are Doing Amazing Things"; "The District Line,"; White, "Changing Rose."

13. For amateur poetry, see McCaig, "Space Seedsman." I dicuss amateur experimentation later in the chapter.

14. "Atomic Garden—a Home Show 'First.'"

15. "Your Results Will Be Shown."

16. Bartell, "Be an 'Atomic Gardener'"; "Free Irradiated Seeds." This effort, though driven by commercial ambitions, resembled the pitch made by O. J. Eigsti in his recruitment of participants to the study of colchicine effects and to some contemporary citizen science projects. See discussion in chapter 9.

17. "Atomic Seeds Are Going Fast"; Bartell, "Thousands Join 'Atomic Seed' Test."

18. E.g., Bartell, "Let Winds Blow"; "Students Check 'Atomic Seeds'"; "Scientists Keep an Eye on 'Atomic Seeds.'"

19. E.g., "Grow a Mystery Plant" [advertisement], E26.

20. Bartell, "Atomic Garden Is New Star."

21. Johns, "Home Show Opens," O1.

22. Kane, "Home Show Opens," 7; Kendall, "'Hot' Seeds."

23. "Home Show Featured in Space Age Garden," 16.

24. E.g., Newman, "New Aids for Garden."

25. "Burpee Has Used Atomic Energy."

26. Breck's atomic varieties may have been purchased from Oak Ridge Atom Industries, as it advertised a subset of varieties also produced by the Oak Ridge firm. See, e.g., "Breck's New Atomic Seeds!" [advertisement].

27. Roessner, "Atomic Age Hits the Garden," B12.

28. "Grow a Mystery Plant" [advertisement], E26.

29. "200 Attend Lecture," 17.

30. "Atomic-Energized Seeds and Plants" [advertisement], 19.

31. Boller, "Unpredictable, Fantastic Flowers," 5; see also Aronson, "Weeders Guide."

32. Mandeville, "'Atomic' Seeds," 23.

33. Taloumis, "Atomic Gardening," 20B.

34. E.g., "Irradiated Bulbs Add Suspense."

35. "Atomic Tomato Developed," 6.

36. E.g., "Atomic-Energized Seeds and Plants" [advertisement], 19; "National Food Stores, 'Atomic Gardening'" [advertisement], B14.

37. Orr, "Home Garden," E4.

38. Kendall, "'Hot' Seeds," 2A.

39. Republished in "Purple Tomato," 10.

40. "District Line," B18.

41. "Spyglass," B-1.

42. "Home Show Opens Here," 11.

43. These letters are gathered in Bartell, "Several Unusual Plants."

44. Although the fad of atomic gardening died out, atomic-energized seeds did leave a lasting legacy as the inspiration for the fictional character Tillie's science fair project in the American playwright Paul Zindel's *The Effect of Gamma Rays on Man-in-the-Moon Marigolds.* The play ran on Broadway in 1971 and won the Pulitzer Prize for Drama that year. It was subsequently adapted for a film, directed and produced by Paul Newman. See Zindel, *Effect of Gamma Rays.*

45. "About Problem Lawns . . ." [advertisement], C15; "Amazing Energized Golf Ball" [advertisement], 2.

46. Clarence J. Speas and Paul L. Andrews, assignors to Oak Ridge Atom Industries, A Method of Treating Grape Vines, US Patent 3,104,497, filed 25 August 1961, and issued 24 September 1963. See also "Patents of the Week."

47. "Meet Dr. C. J. Speas" [advertisement], F10; "Pepped-up Peanut," F14.

48. E.g., "Free Irradiated Seeds."

49. For descriptions of this peanut breeding effort, see Gregory, "Useful Mutations"; Gregory, "X-Ray Breeding."

50. "Amazing NC-4X Peanut Plant" [advertisement], E26.

51. Taloumis, "Atomic Gardening," 20B.

52. Bartell, "Be an 'Atomic Gardener,'" G1.

53. "National Food Stores, 'Atomic Gardening'" [advertisement], B14.

54. Singleton to Wittmeyer, 31 January 1961, Singleton Papers, Box 11; Singleton, "Irradiated Seed," X23.

Chapter Fifteen

1. On the United States and EURATOM, see Hewlett and Holl, *Atoms for Peace and War*, chs. 9 and 16. There were of course many more specific political objectives subsumed within these broad aims. See Medhurst, "Atoms for Peace"; Krige, "Atoms for Peace."

2. Medhurst, "Atoms for Peace," 586-89; Krige, "Atoms for Peace," 172-74, 180-81.

3. On the relationship between nation-building and nuclear capacity, see (on South Korea) DiMoia, "Atoms for Sale"; (on India) Anderson, *Nucleus and Nation*. On the importance accorded to being a "nuclear" state, see Hecht, *Being Nuclear*.

4. An account of long-term global agricultural development (which includes an analysis of post–World War II changes) is Tauger, *Agriculture in World History*.

5. Of necessity, my account of international interest in mutation breeding in the 1950s and 1960s is brief and told from the American perspective; with the exception of Karin Zachmann's research on "nuclear agriculture" in Germany, significant research on mutation breeding in other national contexts and on the global spread of mutation breeding techniques has yet to be conducted. Zachmann's account indicates a trajectory similar to that of the American context in which interest in the use of radiation in plant breeding was revived—and politicized—after World War II in both West and East Germany. See Zachmann, "Peaceful Atoms in Agriculture and Food." Other published accounts of postwar mutation breeding include Hamblin, "Let There Be Light"; Hamblin, "Quickening Nature's Pulse"; Zachmann, "Risky Rays."

6. Hewlett and Holl, *Atoms for Peace and War*, 209–32.

7. There were other goals, too. John Krige discusses the conference aims as also including "scientific intelligence gathering" in which delegates from the United States were expected to press scientists from abroad for technical information. See Krige, "Atoms for Peace," 166. See also Hewlett and Holl, *Atoms for Peace and War*, 232–35.

8. Comments of Silow in "Record of Proceedings of Session 7.2," in *Proceedings of the International Conference on the Peaceful Uses of Atomic Energy* [hereafter *Proceedings*], 12:19–21. Silow later became an ardent critic of the IAEA's promotion of atomic agriculture in developing countries. For further background on Silow and his role at the FAO, see Hamblin, "Let There Be Light"; and Zachmann, "Risky Rays."

9. See contributions to Session 13C.2, "Radiation-Induced Genetic Changes and Crop Improvement," in *Proceedings*, 12:25–71.

10. Comments by Gustafsson, "Record of Proceedings of Session 13C.2," in *Proceedings*, 12:71. Emphasis in original.

11. Ibid., 71.

12. For a discussion of human-caused evolution in insects, weeds, and other organisms, and its costs, see Palumbi, "World's Greatest Evolutionary Force."

13. Smith, "Radiation in the Production of Useful Mutations," 3.

14. Other instances of this argument include, for example, BNL, *Annual Report, July 1, 1954*, 51–52; Libby, "Economic Potential," 5; US Congress, Joint Committee on Atomic Energy, *Progress Report*, 9.

15. Singleton, "Mutation Breeding," talk at Tenth Annual Hybrid Corn Industry-Research Conference, 1955, Singleton Papers, Box 18.

16. FAO, "The Uses of Atomic Energy in Food and Agriculture," in *Proceedings*, 12:19. See discussion in Zachmann, "Risky Rays," 30–33.

17. Hillaby, "Atom May Unlock Arctic," 6.

18. "Nuclear Age Means More Food," 8.

19. For a history of the Green Revolution that gives attention to the political economies that shaped its trajectory, see Perkins, *Geopolitics and the Green Revolution*. See also Cullather, *Hungry World*.

20. A survey of the history of post-1945 interest in "food security" is Shaw, *World Food Security*.

21. On the expansion of mutation breeding and nuclear agriculture, see Hamblin, "Let There Be Light"; Hamblin, "Quickening Nature's Pulse"; Zachmann, "Risky Rays"; Zachmann, "Peaceful Atoms in Agriculture and Food." New agricultural institutions established in this period include the UN Food and Agriculture Organization (1945) and philanthropy-funded research centers such as the International Rice Research Institute (Philippines, 1960), the International Maize and Wheat Improvement Center (Mexico City, 1966), and the International Institute of Tropical Agriculture (Nigeria, 1967). For an overview of national and international agricultural research, especially the development of institutions since World War II, see Tribe, *Feeding and Greening*, ch. 7.

22. Brookhaven's role in the spread of mutation breeding techniques was recognized at the time. See, e.g., Silow, "Potential Contribution of Atomic Energy," 266.

23. Collaborating researchers in 1956 hailed from Australia, Canada, Italy, Mexico, Pakistan, Peru, the Philippines, Romania, Thailand, and Venezuela. See BNL, *Annual Report, July 1, 1956*, 58.

24. Lang, "Stroll in the Garden," 31.

25. Warden, "Chinese Seeds," 4.

26. BNL, *Annual Report, July 1, 1956*, 58.

27. Boroughs, "Radiation Research at a Tropical Center," 25.

28. Moh and Orbegoso, "Effects of Ionizing Radiations."

29. D'Amato et al., "Gamma Radiation Field," 244–45.

30. Kawara, "Introduction of a Gamma Field," 175 (quotations), 176–77.

31. See the partial list of sites hosting gamma fields by the mid-1960s in Woodley, "Open Irradiation Facilities," 89–94.

32. Caldecott, Stevens, and Roberts, "Stem Rust Resistant Variants." See also Griffiths and Johnston, "Irradiation Technique in Oat Breeding."

33. Sagawa and Mehlquist, "X-Ray Induced Changes."

34. E.g., O'Mara to Sears, 18 January 1954, Sears Papers, Folder 550. See discussion in chapter 12.

35. Myers quotation from "Discussion" in MacKey, "Mutation Breeding in Europe," 152.

36. Dick, *Atomic Energy in Agriculture*, 2.

37. Singleton, "Use of Radiation in Plant Breeding," 7.

38. For a short description of the FAO/IAEA program, see van Harten, *Mutation Breeding*, 61–63.

39. "Radiation and the Green Revolution," 18. See also Ahloowalia, Maluszynski, and Nichterlein, "Global Impact," 193.

40. Hamblin, "Quickening Nature's Pulse."

41. "Radiation and the Green Revolution," 18.

42. On the role of atomic agencies in the promotion of mutation breeding in East and West Germany, see Zachmann, "Peaceful Atoms in Agriculture and Food."

43. E.g., Konzak, "Genetic Effects of Radiation," 27; Silow, "Potential Contribution of Atomic," 266; Sparrow and Singleton, "Radiocobalt as a Source of Gamma Rays," 29.

44. Examples of researchers aspiring to the development of precision techniques include Caldecott, "Ionizing Radiations as a Tool for Plant Breeders," in *Proceedings*, 12:40; Konzak, "Genetic Effects of Radiation," 27.

45. Smith, "Radiation in the Production of Useful Mutations," 3.

46. Hamblin, "Quickening Nature's Pulse."

47. Singleton, "Atomic Energy and Agriculture," 2.

48. See, e.g., the "Mutant Variety Database" maintained by the FAO/IAEA, which catalogs induced-mutation varieties developed around the world, http://mvgs.iaea.org/AboutMutant Varities.aspx (accessed 24 August 2014).

49. "Atomic Rays Are Doing Amazing Things," 10F.

Epilogue

1. Hotchkiss quoted in Kotulak, "Your Child Built to Order," 5.

2. For an overview of molecular biology, see Morange, *Molecular Biology*; see also de Chadarevian, *Designs for Life*.

3. Reinhold, "Evolution Control," 1.

4. James, "Genetic Blueprints," 18. On patterns in popular and/or public discourse about genetics and genetic technologies over the course of the twentieth century, see van Dijck, *Imagenation*; Turney, *Frankenstein's Footsteps*; Condit, *Meanings of the Gene*. A detailed survey of (post-1970 only) reporting on biotechnology in the American press, which includes an overview of changing trends in the tone and content of media accounts, is Nisbet and Lewenstein, "Biotechnology and the American Media."

5. E.g., Dighton, "Genetic Tinkering"; "Controlling the Mind." For one biologist's despair over the use of the term "tampering" in reference to molecular biology, see Lederberg, "Genetic Intervention," A13.

6. An account of these events is given in Beckwith's autobiography, *Making Genes*.

7. Shearer, "Change the Human Race," D6.

8. For an overview of the history of in vitro fertilization and other reproductive technologies, see Henig, *Pandora's Baby*.

9. For Berg's recent reflections on this conference (known as the Asilomar Conference on Recombinant DNA), see Berg, "Meetings That Changed the World." For a historical analysis of scientists' intentions in organizing this meeting, and its consequences, see Wright, "Legitimating Genetic Engineering." A full account of the early recombinant DNA controversies is Krimsky, *Genetic Alchemy*.

10. Vellucci quoted in Kifner, "'Creation of Life' Experiment," 22. For an overview of the public hearing held in Cambridge, Massachusetts, in response to a proposed recombinant DNA research laboratory at Harvard, see Goodell, "Public Involvement"; see also Krimsky, *Genetic Alchemy*, ch. 22.

11. This history is the subject of Charles, *Lords of the Harvest*; see also Lurquin, *Green Phoenix*.

12. Hilts, "Test Tube Babies of Agriculture," A1.

13. An account of the controversies over genetically modified crops in the United States is Winston, *Travels*; see also Krimsky, *Biotechnics and Society*, esp. chs. 5–8.

14. May, "Battle Over Plant Gene Splicing," D1.

15. Sheldon Krimsky discusses controversies about the "naturalness" of transgenic manipulation and especially the idea of crossing species boundaries in *Genetic Alchemy*, ch. 19. A few scholars have considered the importance attributed to "naturalness" in debates over genetic technologies; a study focused on the British context is Hansen, "Tampering with Nature." A review of this literature is found in Dragojlovic and Einsiedel, "Framing Synthetic Biology."

16. For overviews of the environmental movement in the United States, see Hays, *Beauty, Health, and Permanence*; Sale, *Green Revolution*; Shabecoff, *Fierce Green Fire*; Gottlieb, *Forcing the Spring*.

17. Shulman, *Seventies*; Berkowitz, *Something Happened*.

18. Mark Winston touches on some of these issues in *Travels*. A global perspective on the political battles over genetically modified crops is offered in Schurman and Munro, *Future of Food*.

19. These are commonly cited in news articles as well as official accounts of mutation breeding. See, e.g., Broad, "Useful Mutants, Bred with Radiation"; van Harten, *Mutation Breeding*, 238, 285; Ahloowalia and Maluszynski, "Induced Mutations," 168–69.

20. Data are from the USDA's Economic Research Service: http://www.ers.usda.gov/data-products/adoption-of-genetically-engineered-crops-in-the-us.aspx (accessed 8 August 2015). A 2004 survey of mutation-bred varieties gives a sense of the scale at which these are cultivated; see Ahloowalia, Maluszynski, and Nichterlein, "Global Impact."

21. Moore, *Disrupting Science*.

22. For a history of these developments in relation to a single biotech start-up, see Hughes, *Genentech*. It is important to note that this speculation occurred primarily in relation to biomedical and not agricultural biotechnologies.

23. The most significant exceptions are the researchers at GE discussed in chapter 5; Albert Blakeslee, who worked at the privately funded Carnegie Institution of Washington's Cold Spring Harbor Station for Experimental Evolution (later Department of Genetics); and David Burpee of W. Atlee Burpee & Co.

24. This point of comparison between mutation technologies and current tools of genetic modification is also made in Murphy, *Plant Breeding*.

25. Boyd, "Wonderful Potencies." See also McAfee, "Biotech Battles."

26. A technical account of glyphosate-resistant crops by a Monsanto employee is Dill, "Glyphosate-Resistant Crops." This development is also discussed in Charles, *Lords of the Harvest*.

27. On the evolution of resistance, see Palumbi, "World's Greatest Evolutionary Force." On chemicals and the problem of pest resistance from a historical perspective, see also Russell, *War and Nature*; Buhs, *Fire Ant Wars*.

28. There is no comprehensive account of such research activities. For some statements of biotechnologists working (or supporting work) in this vein, see Chrispeels, "Biotechnology and the Poor"; Potrykus, "Golden Rice and Beyond."

29. For example, Greenpeace protestors have long objected to the development of Golden Rice, a variety in which the grain carries vitamin A and which was intended as a means of delivering this to children and adults in poorer countries who suffer from vitamin deficiency. For a brief history, see Enserink, "Tough Lessons."

30. Stroman and Lewis, "Genetic Effects of Cosmic Radiation."

31. For a recent report of such research, see Liu et al., "Crop Space Breeding in China."

32. Venter quoted in "Researchers Create the World's First Fully Synthetic, Self-Replicating Living Cell." For more on this research, see Gibson et al., "Creation of a Bacterial Cell."

Bibliography

Many of the newspaper articles cited below can be found in online databases, which are indicated as follows:

Google News Archive   Google News Historical Newspaper Archive,
            https://news.google.com/newspapers
NewspaperArchive   NewspaperArchive, http://newspaperarchive.com
Plain Dealer Archive   Cleveland Plain Dealer Historical Archive,
            http://nl.newsbank.com/nl-search/we/Archives
ProQuest   ProQuest Historical Newspapers, http://search.proquest.com
Utah Digital Newspapers   Utah Digital Newspapers, http://udn.lib.utah.edu

"A-Bomb's Rays Produce Glowing Corn." *Life*, 6 August 1951, 65.
"About Problem Lawns . . ." [advertisement]. *Chicago Daily Tribune*, 11 May 1962, C15. ProQuest.
Adams, Michael C. C. *Best War Ever: America and World War II*. Baltimore: John Hopkins University Press, 1994.
Aftalion, Fred. *A History of the International Chemical Industry: From the "Early Days" to 2000*. Translated by Otto Theodor Benfey. Philadelphia: Chemical Heritage Press, 2001.
Ahloowalia, B. S., and M. Maluszynski. "Induced Mutations—a New Paradigm in Plant Breeding." *Euphytica* 118, no. 2 (2001): 167–73.
Ahloowalia, B. S., M. Maluszynski, and K. Nichterlein. "Global Impact of Mutation-Derived Varieties." *Euphytica* 135, no. 2 (2004): 187–204.
Alexander, Jennifer Karns. *The Mantra of Efficiency: From Waterwheel to Social Control*. Baltimore: Johns Hopkins University Press, 2008.
Allen, Garland E. "The Eugenics Record Office at Cold Spring Harbor, 1910–1940: An Essay in Institutional History." *Osiris* 2 (1986): 225–64.
———. "Hugo de Vries and the Reception of the 'Mutation Theory.'" *Journal of the History of Biology* 2, no. 1 (1969): 55–87.
———. "The Reception of Mendelism in the United States, 1900–1930." *Comptes Rendus de l'Académie des Sciences—Series III—Sciences de la Vie* 323, no. 12 (2000): 1081–88.
———. "Thomas Hunt Morgan and the Problem of Natural Selection." *Journal of the History of Biology* 1, no. 1 (1968): 113–39.

"Amazing Energized Golf Ball Will Improve Your Game" [advertisement]. *Wall Street Journal*, 13 September 1963, 2. ProQuest.

"Amazing NC-4X Peanut Plant" [advertisement]. *Cleveland Plain Dealer*, 20 November 1960, E26. Plain Dealer Archive.

Amundson, Michael A. "'Uranium on the Cranium': Uranium Mining and Popular Culture." In *Atomic Culture: How We Learned to Stop Worrying and Love the Bomb*, edited by Scott C. Zeman and Michael A. Amundson, 49–63. Boulder: University of Colorado Press, 2004.

"Ancient Gout Treatment is Aid to Plant [*sic*]." *Schenectady Gazette*, 11 July 1938, 9. Google News Archive.

Andersen, A. L. "The Sanilac Story." In *Mutation Breeding Workshop, January 17–18, 1972, University of Tennessee Agricultural Campus, Knoxville, Tennessee.* Accessed 14 November 2015. http://www.osti.gov/scitech/biblio/4297499.

Anderson, J. L. *Industrializing the Corn Belt: Agriculture, Technology, and Environment, 1945–1972.* DeKalb: Northern Illinois University Press, 2008.

———. "War on Weeds: Iowa Farmers and Growth-Regulator Herbicides." *Technology and Culture* 46, no. 4 (2005): 719–44.

Anderson, Robert S. *Nucleus and Nation: Scientists, International Networks, and Power in India.* Chicago: University Chicago Press, 2010.

Apple, Rima D. *Vitamania: Vitamins in American Culture.* New Brunswick, NJ: Rutgers University Press, 1996.

Aronson, Earl. "The Weeders Guide: Atomic Irradiated Seeds Produce Striking Plants." *Hartford Courant*, 12 November 1960, 5. ProQuest.

"Atom Alters Heredity." *Science News-Letter* 58, no. 12 (1950): 188.

"Atomic-Energized Seeds and Plants" [advertisement]. *Chicago Tribune*, 14 April 1961, 19. ProQuest.

"Atomic Garden—a Home Show 'First.'" *Cleveland Plain Dealer*, 18 December 1960, H10. Plain Dealer Archive.

"Atomic Oven Bakes Seeds to Aid Crops." *Baltimore Sun*, 28 March 1949, 3. ProQuest.

"Atomic Rays Are Doing Amazing Things to Seeds." *Hartford Courant*, 10 April 1960, 10F.

"Atomic Seeds Are Going Fast." *Cleveland Plain Dealer*, 15 November 1960, A1. Plain Dealer Archive.

"Atomic Tomato Developed." *Lodi News-Sentinel*, 12 June 1962, 6. Google News Archive.

"Atom Ray Used on Seeds to Improve Crops." *Los Angeles Times*, 28 March 1949, 17. ProQuest.

"Atom Study Points to Food Plenty by Fast Development of New Plants." *New York Times*, 31 January 1952, 4. ProQuest.

Babcock, E. B. "John Belling: Pioneer in the Study of Cell Mechanics." *Journal of Heredity* 24, no. 8 (1933): 297–300.

Babcock, Ernest Brown, and Roy Elwood Clausen. *Genetics in Relation to Agriculture.* New York: McGraw-Hill, 1927.

Baker, Herbert G., Adriance S. Foster, and Johannes M. Proskauer. "Thomas Harper Goodspeed, Botany: Berkeley." In *University of California: In Memoriam 1967*, edited by University of California Academic Senate, 43–46. Berkeley: University of California, 1967. Accessed 4 July 2015. http://texts.cdlib.org/view?docId=hb629006vt.

Baker, Richard Elwin. "Induced Polyploid, Periclinal Chimeras in Solanum Tuberosum." *American Journal of Botany* 30, no. 3 (1943): 187–95.

Bartell, Irma. "Atomic Garden Is New Star in 9-Day Hit Show." *Cleveland Plain Dealer*, 5 March 1961, F1-2. Plain Dealer Archive.

——. "Atomic Seeds Produce Several Unusual Plants." *Cleveland Plain Dealer*, 12 November 1961, G10. Plain Dealer Archive.

——. "Be an 'Atomic Gardener' and Join a Team of Scientific Pioneers." *Cleveland Plain Dealer*, 13 November 1960, G1, 2. Plain Dealer Archive.

——. "Let Winds Blow! Your Greenhouse Is Tropical Haven." *Cleveland Plain Dealer*, 27 November 1960, H15. Plain Dealer Archive.

——. "Thousands Join 'Atomic Seed' Test, but It Takes Time." *Cleveland Plain Dealer*, 20 November 1960, H14. Plain Dealer Archive.

Bashford, Alison, and Philippa Levine, eds. *The Oxford Handbook of the History of Eugenics*. Oxford: Oxford University Press, 2010.

Bates, G. H. "Polyploidy Induced by Colchicine and Its Economic Possibilities." *Nature* 144, no. 3642 (1939): 315–16.

Beatty, John. "Genetics in the Atomic Age: The Atomic Bomb Casualty Commission, 1947–1956." In *The Expansion of American Biology*, edited by Keith R. Benson, Jane Maienschein, and Ronald Rainger, 284–324. New Brunswick, NJ: Rutgers University Press, 1991.

——. "Masking Disagreement among Experts." *Episteme* 3, no. 1–2 (2006): 52–67.

——. "Scientific Collaboration, Internationalism, and Diplomacy: The Case of the Atomic Bomb Casualty Commission." *Journal of the History of Biology* 26, no. 2 (1993): 205–31.

——. "Weighing the Risks: Stalemate in the Classical/Balance Controversy." *Journal of the History of Biology* 20, no. 3 (1987): 289–319.

Beckwith, Jon. *Making Genes, Making Waves: A Social Activist in Science*. Cambridge, MA: Harvard University Press, 2002.

Beckwith, T. D., A. R. Olson, and E. J. Rose. "The Effect of X-Ray upon Bacteriophage and upon the Bacterial Organism." *Proceedings of the Society for Experimental Biology and Medicine* 27, no. 4 (1930): 285–86.

Beeman, Randal. "'Chemivisions': The Forgotten Promises of the Chemurgy Movement." *Agricultural History* 68, no. 4 (1994): 23–45.

Belling, John. "The Iron-Acetocarmine Method of Fixing and Staining Chromosomes." *Biological Bulletin* 50, no. 2 (1926): 160–62.

——. "Production of Triploid and Tetraploid Plants." *Journal of Heredity* 16, no. 12 (1925): 463–64.

Benson, Keith R. "From Museum Research to Laboratory Research: The Transformation of Natural History into Academic Biology." In *The American Development of Biology*, edited by Ronald Rainger, Keith R. Benson, and Jane Maienschein, 49–83. Philadelphia: University of Pennsylvania Press, 1988.

Bentley, Amy. *Eating for Victory: Food Rationing and the Politics of Domesticity*. Urbana: University of Illinois Press, 1998.

Berg, Paul. "Meetings That Changed the World: Asilomar 1975: DNA Modification Secured." *Nature* 455, no. 7211 (2008): 290–91.

Berkowitz, Edward D. *Something Happened: A Political and Cultural Overview of the Seventies*. New York: Columbia University Press, 2006.

"Better Regal Lily Produced by X-Ray." *New York Times*, 1 September 1935, N1. ProQuest.

"Bigger and Better Berries Fruit Trees and Flowers." *Albuquerque Journal*, 3 March 1940, Magazine 6. NewspaperArchive.

Bijker, Wiebe E. *Of Bicycles, Bakelites, and Bulbs: Toward a Theory of Sociotechnical Change*. Cambridge, MA: MIT Press, 1995.

Bijker, Wiebe E., Thomas P. Hughes, and Trevor Pinch, eds. *The Social Construction of Technological Systems: New Directions in the Sociology and History of Technology*. Anniv. ed. Cambridge, MA: MIT Press, 2012.

Birr, Kendall. *Pioneering in Industrial Research: The Story of the General Electric Research Laboratory*. Washington, DC: Public Affairs Press, 1957.

Bishop, C. J. "Mutations in Apples Induced by X-Radiation." *Journal of Heredity* 45, no. 2 (1954): 99–104.

Blair, William M. "Strauss Praises Atomic Crop Gain." *New York Times*, 20 September 1957, 23. ProQuest.

Blakeslee, A. F. "Annual Report of the Director of the Department of Genetics." In *Carnegie Institution of Washington Year Book No. 38 (1938–1939)*, 176–208. Washington, DC: Carnegie Institution of Washington, 1939.

———. "Department of Genetics, Chromosome Investigations." In *Carnegie Institution of Washington Year Book No. 37 (1937–1938)*, 35–40. Washington, DC: Carnegie Institution of Washington, 1938.

———. "John Belling: October 7, 1866—February 16, 1933." *Stain Technology* 8, no. 3 (1933): 83–86.

———. "New Jimson Weeds from Old Chromosomes." *Journal of Heredity* 25, no. 3 (1934): 81–108.

———. "Types of Mutations and Their Possible Significance in Evolution." *American Naturalist* 55, no. 638 (1921): 254–67.

———. "Variations in Datura Due to Changes in Chromosome Number." *American Naturalist* 56, no. 642 (1922): 16–31.

Blakeslee, Albert F., and Amos G. Avery. "Methods of Inducing Doubling of Chromosomes in Plants by Treatment with Colchicine." *Journal of Heredity* 28, no. 12 (1937): 393–411.

Blakeslee, A. F., and B. T. Avery. "Adzuki Beans and Jimson Weeds: Favorable Class Material for Illustrating the Ratios of Mendel's Law." *Journal of Heredity* 8, no. 3 (1917): 125–31.

———. "Mutations in the Jimson Weed." *Journal of Heredity* 10, no. 3 (1919): 111–20.

Blakeslee, A. F., and John Belling. "Chromosomal Mutations in the Jimson Weed, Datura Stramonium." *Journal of Heredity* 15, no. 5 (1924): 195–206.

Blakeslee, Albert F., John Belling, and M. E. Farnham. "Chromosomal Duplication and Mendelian Phenomena in Datura Mutants." *Science* 52, no. 1347 (1920): 388–90.

Blanchard, G. A. "Colchicine in Gladiolous Breeding." *Horticulture*, 1 March 1940, 100.

Bliven, Bruce. "Remaking the World of Plants." *New Republic*, 12 May 1941, 656–59.

Bocking, Stephen. *Ecologists and Environmental Politics: A History of Contemporary Ecology*. New Haven, CT: Yale University Press, 1997.

———. "Ecosystems, Ecologists, and the Atom: Environmental Research at Oak Ridge National Laboratory." *Journal of the History of Biology* 28, no. 1 (1995): 1–47.

Boersma, F. Kees. "The Organization of Industrial Research as a Network Activity: Agricultural Research at Philips in the 1930s." *Business History Review* 78, no. 2 (2004): 255–72.

———. "Structural Ways to Embed a Research Laboratory into the Company: A Comparison between Philips and General Electric 1900–1940." *History and Technology* 19, no. 2 (2003): 109–26.

Boller, Fay. "Unpredictable, Fantastic Flowers and Vegetables Produced by Seeds Treated with Cobalt-60." *Toledo Blade*, 1 April 1961, 5. Google News Archive.

Borlaug, Norman E. "Ending World Hunger. The Promise of Biotechnology and the Threat of Antiscience Zealotry." *Plant Physiology* 124, no. 2 (2000): 487–90.

Boroughs, Howard. "Radiation Research at a Tropical Center." *AIBS Bulletin* 11, no. 6 (1961): 25–26.

Borup, Mads, Nik Brown, Kornelia Konrad, and Harro Van Lente. "The Sociology of Expectations in Science and Technology." *Technology Analysis & Strategic Management* 18, no. 3/4 (2006): 285–98.

Bott, W. E. "Test Tube Garden" [3-part series]. *Cleveland Press*, 22–24 June 1939. Accessed 14 November 2015. http://w.lakewoodhistory.org/history%20files/5biographyA-F.htm.

Bowler, Peter J. *Evolution: The History of an Idea*. 25th anniv. ed. Berkeley: University of California Press, 2009.

———. "Hugo De Vries and Thomas Hunt Morgan: The Mutation Theory and the Spirit of Darwinism." *Annals of Science* 35, no. 1 (1978): 55–73.

———. *The Mendelian Revolution: The Emergence of Hereditarian Concepts in Modern Science and Society*. London: Athlone, 1989.

Boyd, William. "Wonderful Potencies? Deep Structure and the Problem of Monopoly in Agricultural Biotechnology." In *Engineering Trouble: Biotechnology and Its Discontents*, edited by Rachel A. Schurman and Dennis Doyle Takahashi Kelso, 24–62. Berkeley: University of California Press, 2003.

Boyer, Paul. *By the Bomb's Early Light: American Thought and Culture at the Dawn of the Atomic Age*. New York: Pantheon, 1985.

"Breck's New Atomic Seeds!" [advertisement]. *New York Times*, 19 March 1961, X24. ProQuest.

Brewster, E. T. "Breeding Plants and Animals to Order." *World's Work*, December 1907, 9653–58.

Bridges, Calvin B. "A Linkage Variation in Drosophila." *Journal of Experimental Zoology* 19, no. 1 (1915): 1–21.

Broad, William J. "Useful Mutants, Bred with Radiation." *New York Times*, 28 August 2007. Accessed 2 August 2013. http://www.nytimes.com/2007/08/28/science/28crop.html.

Brookhaven National Laboratory. *Annual Report, July 1, 1950*. Upton, NY: Associated Universities Inc., 1950.

———. *Annual Report, July 1, 1952*. Upton, NY: Associated Universities Inc., 1952.

———. *Annual Report, July 1, 1953*. Upton, NY: Associated Universities Inc., 1953.

———. *Annual Report, July 1, 1954*. Upton, NY: Associated Universities Inc., 1954.

———. *Annual Report, July 1, 1956*. Upton, NY: Associated Universities Inc., 1956.

———. *Conference on Biological Applications of Nuclear Physics, July 12–27, 1948*. Upton, NY: Brookhaven National Laboratory, 1948.

Brown, Nik. "Hope against Hype—Accountability in Biopasts, Presents, and Futures." *Science Studies* 16, no. 2 (2003): 3–21.

Brownell, Frederick G. *The Seeds of Tomorrow: Celebrating One Hundred Years of Service to Commercial Growers and Home Gardeners, 1856–1956*. N.p.: Ferry-Morse Seed Co., 1956.

Buchanan, Nicholas. "The Atomic Meal: The Cold War and Irradiated Foods, 1945–1963." *History and Technology* 21, no. 2 (2005): 221–49.

Bud, Robert. *The Uses of Life: A History of Biotechnology*. Cambridge: Cambridge University Press, 1993.

Bugos, Glenn E., and Daniel J. Kevles. "Plants as Intellectual Property: American Practice, Law, and Policy in World Context." *Osiris* 7 (1992): 75–104.

Buhs, Joshua Blu. *The Fire Ant Wars: Nature, Science, and Public Policy in Twentieth-Century America*. Chicago: University of Chicago Press, 2004.

Burpee, David, and Frank J. Taylor. "So We Shocked Mother Nature." *Rotarian*, July 1940, 15–17.

"Burpee for Burbank." *Time*, 21 September 1931. Accessed 24 February 2012. http://www.time.com/time/magazine/article/ 0,9171,742297,00.html.

"Burpee Has Used Atomic Energy to Increase Your Chances to Grow a White Marigold—and Get $10,000" [advertisement]. *Pittsburgh Press*, 7 January 1962, sec. 6, 10. Google News Archive.

"Burpee Introduces His New Calendulas." *New York Times*, 27 January 1942, 29. ProQuest.

Caldecott, Richard S., Harland Stevens, and Bill J. Roberts. "Stem Rust Resistant Variants in Irradiated Populations: Mutations or Field Hybrids?" *Agronomy Journal* 51, no. 7 (1959): 401–3.

California Agricultural Experiment Station. *Report of the Agricultural Experiment Station of the University of California, from July 1, 1927, to June 30, 1928, Part 1.* Berkeley: University of California Printing Office, 1929.

——. *Report of the Agricultural Experiment Station of the University of California, from July 1, 1928, to June 30, 1929.* Berkeley: University of California Printing Office, 1930.

Callaway, Dorothy J. *The World of Magnolias.* Portland, OR: Timber Press, 1994.

"Cambridge Laboratories" [advertisement]. *Science News-Letter* 37, no. 3 (1940): 43.

Campos, Luis. "The Birth of Living Radium." *Representations* 97, no. 1 (2007): 1–27.

——. "Genetics without Genes: Blakeslee, Datura, and 'Chromosomal Mutations.'" In "A Cultural History of Heredity IV: Heredity in the Century of the Gene," 243–57. Preprint 343. Berlin: Max Planck Institute for the History of Science, 2008.

——. *Radium and the Secret of Life.* Chicago: University of Chicago Press, 2015.

——. "That Was the Synthetic Biology That Was." In *Synthetic Biology: The Technoscience and Its Societal Consequences*, edited by Markus Schmidt, Alexander Kelle, Agomoni Ganguli-Mitra, and Huib de Vriend, 5–21. Dordrecht, Neth.: Springer, 2009.

Campos, Luis, and Alexander von Schwerin, eds. "Making Mutations: Objects, Practices, Contexts." Preprint 393. Berlin: Max Planck Institute for the History of Science, 2010.

"Can A-Bomb Breed 4-Inch Monsters?" *Los Angeles Times*, 16 December 1951, 35. ProQuest.

Carlson, Elof Axel. *Genes, Radiation, and Society: The Life and Work of H. J. Muller.* Ithaca, NY: Cornell University Press, 1981.

"Cathode Ray Yields Chemical Secrets." *New York Times*, 16 June 1932, 8. ProQuest.

Caufield, Catherine. *Multiple Exposures: Chronicles of the Radiation Age.* Chicago: University of Chicago Press, 1989.

Chandler, Alfred D., Jr. *Shaping the Industrial Century: The Remarkable Story of the Evolution of the Modern Chemical and Pharmaceutical Industries.* Cambridge, MA: Harvard University Press, 2005.

Charles, Daniel. *Lords of the Harvest: Biotech, Big Money, and the Future of Food.* Cambridge, MA: Perseus Pub., 2001.

Charnley, Berris, and Gregory Radick. "Intellectual Property, Plant Breeding and the Making of Mendelian Genetics." *Studies in History and Philosophy of Science Part A* 44, no. 2 (2013): 222–33.

"Chemical Born Flower Will Be on View Sunday." *Chicago Daily Tribune*, 10 November 1940, N8. ProQuest.

"Chemical-Created Flower Put on Exhibition Here." *New York Times*, January 30 1940, 6. ProQuest.

"Chemical Is Offered for Botanical Tests." *Fitchburg Sentinel*, 2 July 1940, 6. NewspaperArchive.

Chrispeels, Maarten J. "Biotechnology and the Poor." *Plant Physiology* 124, no. 1 (2000): 3–6.

Clarke, Robert Connell. *Marijuana Botany, an Advanced Study: The Propagation and Breeding of Distinctive Cannabis.* Berkeley, CA: Ronin, 1981.

Clausen, R. E., and T. H. Goodspeed. "Interspecific Hybridization in Nicotiana. II. A Tetraploid *Glutinosa-Tabacum* Hybrid, an Experimental Verification of Winge's Hypothesis." *Genetics* 10, no. 3 (1925): 278–84.

Coe, Ed. "East, Emerson, and the Birth of Maize Genetics." In *Handbook of Maize Genetics*, edited by Jeff Bennetzen and Sarah Hake, 3–15. New York: Springer, 2009.

———. "The Origins of Maize Genetics." *Nature Reviews Genetics* 2, no. 11 (2001): 898–905.

"Colchicine Experimenters Warned of Possible Danger." *Science News-Letter* 38, no. 24 (1940): 382.

Comfort, Nathaniel C. *The Tangled Field: Barbara McClintock's Search for the Patterns of Genetic Control*. Cambridge, MA: Harvard University Press, 2001.

Condit, Celeste Michelle. *The Meanings of the Gene: Public Debates about Human Heredity*. Madison: University of Wisconsin Press, 1999.

Conger, Alan D. "Arnold Hicks Sparrow (1914–1976)." *Radiation Research* 69, no. 1 (1977): 194–96.

Conkin, Paul K. *A Revolution Down on the Farm: The Transformation of American Agriculture since 1929*. Lexington: University of Kentucky Press, 2008.

Considine, Bob. "Behind the Scenes at Brookhaven National Laboratory." *New York Journal American*, 25 September 1949.

"Controlling the Mind." *New York Times*, 29 December 1965, 28. ProQuest.

"Control over Fundamental Life Processes Announced by Eastern Geneticist." *Spokane Daily Chronicle*, 26 October 1937, 20. Google News Archive.

Cooke, Kathy J. "From Science to Practice, or Practice to Science? Chickens and Eggs in Raymond Pearl's Agricultural Breeding Research, 1907–1916." *Isis* 88, no. 1 (1997): 62–86.

Cooley, Donald G. "Apples as Big as Your Head!" *Mechanix Illustrated*, March 1949, 78–81, 145.

Cooper, Melinda. *Life as Surplus: Biotechnology and Capitalism in the Neoliberal Era*. St. Louis, MO: University of Washington Press, 2008.

Cotter, Joseph. *Troubled Harvest: Agronomy and Revolution in Mexico, 1880–2002*. Westport, CT: Praeger, 2003.

Couch, Glenn. "A Botanist Upsets Heredity." *Sooner*, January 1939, 11, 27. Accessed 11 August 2015. https://digital.libraries.ou.edu/sooner/articles/p11,27_1939v11n5_OCR.pdf.

Coulter, John M. *Fundamentals of Plant-Breeding*. New York: D. A. Appleton, 1914.

Cowan, Ruth Schwartz. *A Social History of American Technology*. Oxford: Oxford University Press, 1997.

Creager, Angela N. H. "Biotechnology and Blood: Edwin Cohn's Plasma Fractionation Project, 1940–1953." In *Private Science: Biotechnology and the Rise of the Molecular Sciences*, edited by Arnold Thackray, 39–62. Philadelphia: University of Pennsylvania Press, 1998.

———. "The Industrialization of Radioisotopes by the Atomic Energy Commission." In *The Science-Industry Nexus: History, Policy, Implications*, edited by Karl Grandin, Nina Wormbs, and Sven Widmalm, 141–67. Nobel Symposium 123. Sagamore Beach, MA: Science History Publications/USA, 2004.

———. *Life Atomic: A History of Radioisotopes in Science and Medicine*. Chicago: University of Chicago Press, 2013.

———. "Nuclear Energy in the Service of Biomedicine: The U.S. Atomic Energy Commission's Radioisotope Program, 1946–1950." *Journal of the History of Biology* 39, no. 4 (2006): 649–84.

———. "Phosphorus-32 in the Phage Group: Radioisotopes as Historical Tracers of Molecular Biology." *Studies in History and Philosophy of Science Part C* 40, no. 1 (2009): 29–42.

———. "Radioisotopes as Political Instruments, 1946–1953." *Dynamis* 29 (2009): 219–39.

———. "Tracing the Politics of Changing Postwar Research Practices: The Export of 'American' Radioisotopes to European Biologists." *Studies in History and Philosophy of Science Part C* 33, no. 3 (2002): 367–88.

Creager, Angela N. H., and María Jesús Santesmases. "Radiobiology in the Atomic Age: Changing Research Practices and Policies in Comparative Perspective." *Journal of the History of Biology* 39, no. 4 (2006): 637–47.

Crease, Robert P. *Making Physics: A Biography of Brookhaven National Laboratory, 1946–1972.* Chicago: University of Chicago Press, 1999.

Cresskill, Rita. "Supremes with Colchicine." *African Violet Magazine,* December 1959, 9.

Cronon, William. *Nature's Metropolis: Chicago and the Great West.* New York: W. W. Norton, 1991.

"Crowds 150 Years of Fly Evolution Into 12 Months." *Pittsburgh Press,* 2 July 1930, 38. Google News Archive.

Cullather, Nick. *The Hungry World: America's Cold War Battle against Poverty in Asia.* Cambridge, MA: Harvard University Press, 2010.

Curtis, Howard J. "Leslie F. Nims (1906–1971)." *Radiation Research* 48, no. 2 (1971): 419.

Curtis, Winterton C. "Old Problems and a New Technique." *Science* 67, no. 1728 (1928): 141–49.

Dadourian, Alice B. *A Bio-Bibliography of Caryl Parker Haskins.* Accessed 4 July 2015. http://www.haskins.yale.edu/history/haskinsbio.pdf.

D'Amato, F., G. T. Scarascia, U. Belliazzi, A. Bassani, S. Cambi, P. Cevolotto, P. Giacalone, and S. Tagliati. "The Gamma Radiation Field of the 'Comitato Nazionale Per L'Energia Nucleare,' Rome." *Radiation Botany* 1 (1962): 243–46.

Danbom, David B. *Born in the Country: A History of Rural America.* Baltimore: Johns Hopkins University Press, 1995.

Davenport, C. B. "Annual Report of the Director of the Department of Genetics." In *Carnegie Institution of Washington Year Book No. 21 (1922),* 93–125. Washington, DC: Carnegie Institution of Washington, 1923.

———. "Annual Report of the Director of the Department of Genetics." In *Carnegie Institution of Washington Year Book No. 25 (1926),* 29–59. Washington, DC: Carnegie Institution of Washington, 1926.

———. "First Report of Station for Experimental Evolution under Department of Experimental Biology." In *Carnegie Institution of Washington Year Book No. 3 (1904),* 23–49. Washington, DC: Carnegie Institution of Washington, 1905.

Davey, Wheeler P. "Application of the Coolidge Tube to Metallurgical Research." *General Electric Review* 18, no. 2 (1915): 134–36.

———. "The Effects of X-Rays on the Length of Life of Trilobium Confusum." *General Electric Review* 20, no. 2 (1917): 174–82.

———. "Some Interesting Applications of the Coolidge X-Ray Tube." *General Electric Review* 17, no. 6 (1914): 792–98.

David, Harry M. "New Elixir Found for Plant World." *New York Times,* 16 October 1937, 17. ProQuest.

Davis, Watson. "Magic Wand of X-Ray Found Changing Life." *Pittsburgh Press,* 29 December 1927, 26. Google News Archive.

de Chadarevian, Soraya. *Designs for Life: Molecular Biology after World War II.* Cambridge: Cambridge University Press, 2002.

de Lourdes, Mary. "Growth-Regulating Substances in Plants." *American Biology Teacher* 6, no. 7 (1944): 151–52, 168.

Dennis, Michael Aaron. "Accounting for Research: New Histories of Corporate Laboratories and the Social History of American Science." *Social Studies of Science* 17, no. 3 (1987): 479–518.

Dermen, Haig. "Colchicine Polyploidy and Technique." *Botanical Review* 6, no. 11 (1940): 599–635.

———. "Colchiploidy and Histological Imbalance in Triploid Apple and Pear." *American Journal of Botany* 52, no. 4 (1965): 353–59.

———. "Colchiploidy in Grapes." *Journal of Heredity* 45, no. 4 (1954): 159–72.

Detlefsen, J. A., and L. S. Clemente. "Genetic Variation in Linkage Values." *Proceedings of the National Academy of Sciences of the United States of America* 9, no. 5 (1923): 149–56.

Detlefsen, J. A., and E. Roberts. "Studies on Crossing Over. I. The Effect of Selection on Crossover Values." *Journal of Experimental Zoology* 32, no. 2 (1921): 333–54.

de Vries, Hugo. "Luther Burbank's Ideas on Scientific Horticulture." *Century*, March 1907, 674–81.

———. *Die Mutationstheorie: Versuche und Beobachtungen über die Entstehung von Arten im Pflanzenreich.* 2 vols. Leipzig: Veit, 1901–1903.

———. *The Mutation Theory: Experiments and Observations on the Origin of Species in the Vegetable Kingdom.* Translated by J. B. Farmer and A. D. Darbishire. 2 vols. Chicago: Open Court, 1909–1910.

———. *Plant-Breeding; Comments on the Experiments of Nilsson and Burbank.* Chicago: Open Court, 1907.

———. "A Visit to Luther Burbank." *Popular Science Monthly*, August 1905, 329–47.

Dick, William E. *Atomic Energy in Agriculture.* London: Butterworths Scientific Publication, 1957.

"Dieffenbach, Radiotherapy Specialist Dies." *Neuropath and Herald of Health* 41 (1937): 126.

Dietz, David. "Drug Creates Bigger Fruits, New Flowers." *Pittsburgh Press*, 28 October 1938, 45. Google News Archive.

Dighton, Ralph. "'Genetic Tinkering' Is Term Striking Both Hope and Fear in Men's Mind." *Gettysburg Times*, 12 August 1966, 13. Google News Archive.

Dill, Gerald M. "Glyphosate-Resistant Crops: History, Status and Future." *Pest Management Science* 61, no. 3 (2005): 219–24.

DiMoia, John. "Atoms for Sale? Cold War Institution-Building and the South Korean Atomic Energy Project, 1945–1965." *Technology and Culture* 51, no. 3 (2010): 589–618.

"The District Line: A New Horticultural Roulette Game." *Washington Post and Times Herald*, 17 February 1959, B18. ProQuest.

Dorsey, E. "Induced Polyploidy in Wheat and Rye." *Journal of Heredity* 27, no. 4 (1936): 155–60.

Douglas, Susan J. "Audio Outlaws: Radio and Phonograph Enthusiasts." In *Possible Dreams: Enthusiasm for Technology in America*, edited by John L. Wright, 44–59. Dearborn, MI: Henry Ford Museum & Greenfield Village, 1992.

———. *Inventing American Broadcasting, 1899–1922.* Baltimore: Johns Hopkins Press, 1989.

Dragojlovic, Nick, and Edna Einsiedel. "Framing Synthetic Biology: Evolutionary Distance, Conceptions of Nature, and the Unnaturalness Objection." *Science Communication* 35, no. 5 (2013): 547–71.

Dreyer, Peter. *A Gardener Touched with Genius: The Life of Luther Burbank.* Berkeley: University of California Press, 1985.

Duggar, Benjamin M., ed. *Biological Effects of Radiation: Mechanism and Measurement of Radiation, Applications in Biology, Photochemical Reactions, Effects of Radiant Energy on Organisms and Organic Products.* 2 vols. New York: McGraw-Hill, 1936.

Dunn, L. C. *A Short History of Genetics: The Development of Some of the Main Lines of Thought: 1864–1939.* New York: McGraw-Hill, 1965.

Early, Eleanor. "X-Rays to Produce Giant Flowers, Fruits, and Animals." *Public Ledger Sunday Magazine*, 26 April 1931, 6.

East, E. M. "The Chromosome View of Heredity and Its Meaning to Plant Breeders." *American Naturalist* 49, no. 584 (1915): 457–94.

Edgerton, David. *Shock of the Old: Technology and Global History since 1900.* London: Profile, 2006.

Editorial. "Colchicine a Dangerous Drug." *Journal of Heredity* 29, no. 5 (1938): 188.

———. "Colchicine and Double Diploids." *Journal of Heredity* 28, no. 12 (1937): 411–12.

Egan, Michael. *Barry Commoner and the Science of Survival: The Remaking of American Environmentalism.* Cambridge, MA: MIT Press, 2007.

Eigsti, O. J. *Colchicine Bibliography.* Reprint from *Lloydia,* with supplement by P. Dustin. Cincinnati: Lloyd Library, 1947.

———. "A Cytological Study of Colchicine Effects in the Induction of Polyploidy in Plants." *Proceedings of the National Academy of Sciences of the United States of America* 24, no. 2 (1938): 56–63.

Eigsti, O. J., and Pierre Dustin Jr. *Colchicine in Agriculture, Medicine, Biology and Chemistry.* Ames, IA: State College Press, 1955.

Eigsti, O. J., P. Dustin Jr., and N. Gay-Winn. "On the Discovery of the Action of Colchicine on Mitosis in 1889." *Science* 110, no. 2869 (1949): 692.

Eigsti, O. J., and Barbara Tenney. *A Report on Experiments with Colchicine by Laymen Scientists during 1941.* Norman: University of Oklahoma Press, 1942.

Emsweller, S. L., and M. L. Ruttle. "Induced Polyploidy in Floriculture." *American Naturalist* 75, no. 759 (1941): 310–28.

Endersby, Jim. *A Guinea Pig's History of Biology.* Cambridge, MA: Harvard University Press, 2007.

———. "Mutant Utopias: Evening Primroses and Imagined Futures in Early Twentieth-Century America." *Isis* 104, no. 3 (2013): 471–503.

Enserink, Martin. "Tough Lessons from Golden Rice." *Science* 320, no. 5875 (2008): 468–71.

"Farm Boom Beginning, 4-Club Campers Told." *Washington Post,* 26 June 1928, 10. ProQuest.

"Feathers on Birds Determined by Skin." *New York Times,* 31 December 1927, 6. ProQuest.

Fedoroff, Nina V. "The Past, Present and Future of Crop Genetic Modification." *New Biotechnology* 27, no. 5 (2010): 461–65.

Fedoroff, Nina, and Nancy Marie Brown. *Mendel in the Kitchen: A Scientist's View of Genetically Modified Foods.* Washington, DC: Joseph Henry Press, 2004.

Finlay, Mark R. *Growing American Rubber: Strategic Plants and the Politics of National Security.* New Brunswick, NJ: Rutgers University Press, 2009.

———. "Old Efforts at New Uses: A Brief History of Chemurgy and the American Search for Biobased Materials." *Journal of Industrial Ecology* 7, no. 3–4 (2003): 33–46.

"First X-Ray Flowers to Be Shown Here." *New York Times,* 23 January 1942, 22.

Fitzgerald, Deborah. *The Business of Breeding: Hybrid Corn in Illinois, 1890–1940.* Ithaca, NY: Cornell University Press, 1990.

———. *Every Farm a Factory: The Industrial Ideal in American Agriculture.* New Haven, CT: Yale University Press, 2003.

———. "Exporting American Agriculture: The Rockefeller Foundation in Mexico, 1943–53." *Social Studies of Science* 16, no. 3 (1986): 457–83.

———. "Technology and Agriculture in Twentieth-Century America." In *A Companion to American Technology,* edited by Carroll Pursell, 69–82. Oxford: Blackwell, 2008.

Foster, J. P. "Artificially Induced Mutations." *Planter and Sugar Manufacturer* 82, no. 2 (1929): 21–23.

Fowler, Cary. "The Plant Patent Act of 1930: A Sociological History of Its Creation." *Journal of the Patent and Trademark Office Society* 82, no. 9 (2000): 621–44.

Frank, Glenn. "The No-Man's Land of American Policy." *Century*, March 1919, 635–50.

Franz, Kathleen. *Tinkering: Consumers Reinvent the Early Automobile.* Philadelphia: University of Pennsylvania Press, 2005.

"Free Irradiated Seeds Offered by Plain Dealer." *Cleveland Plain Dealer*, 13 November 1960, G1. Plain Dealer Archive.

Gager, Charles Stuart. *Effects of the Rays of Radium on Plants.* Memoirs of the New York Botanical Garden 4. New York: New York Botanical Garden, 1908.

Gager, C. Stuart, and A. F. Blakeslee. "Chromosome and Gene Mutations in Datura following Exposure to Radium Rays." *Proceedings of the National Academy of Sciences of the United States of America* 13, no. 2 (1927): 75–79.

Galinat, Walton C. "In Memoriam: Willard Ralph Singleton, 1900–1982." *Journal of Heredity* 74, no. 3 (1983): 197–98.

"Garden Time: Clean Bees, Radiation and Roguers Perk Up Seed Growing Business." *Wall Street Journal*, 15 February 1956, 1. ProQuest.

Gardner, Bruce L. *American Agriculture in the Twentieth Century: How It Flourished and What It Cost.* Cambridge, MA: Harvard University Press, 2002.

Gardner, Mona. "Giants in the Garden." *American Magazine*, August 1940, 48, 85–86.

Garner, W. W. "Superior Germ Plasm in Tobacco." In *Yearbook of Agriculture 1936*, 785–830. Washington, DC: USDA, 1936.

Gates, Reginald Ruggles. "The Behavior of the Chromosomes in *Oenothera Lata* × *O. Gigas*." *Botanical Gazette* 48, no. 3 (1909): 179–99.

———. *The Mutation Factor in Evolution, with Particular Reference to Oenothera.* London: MacMillian, 1915.

———. "Mutations and Evolution (Continued)." *New Phytologist* 19, no. 3/4 (1920): 64–88.

———. "Pollen Development in Hybrids of *Oenothera Lata* × *O. Lamarckiana*, and Its Relation to Mutation." *Botanical Gazette* 43, no. 2 (1907): 81–115.

———. "Polyploidy." *Journal of Experimental Biology* 1, no. 2 (1924): 153–82.

Gaudillière, Jean-Paul. "New Wine in Old Bottles? The Biotechnology Problem in the History of Molecular Biology." *Studies in History and Philosophy of Science Part C* 40, no. 1 (2009): 20–28.

———. "Normal Pathways: Controlling Isotopes and Building Biomedical Research in Postwar France." *Journal of the History of Biology* 39, no. 4 (2006): 737–64.

"General Reports of the Second Nashville Meeting of the American Association for the Advancement of Science and Associated Societies." *Science* 67, no. 1726 (1928): 77–85.

Gibson, Daniel G., et al. "Creation of a Bacterial Cell Controlled by a Chemically Synthesized Genome." *Science* 329, no. 5987 (2010): 52–56.

Gibson, G. E., T. D. Stewart, and G. K. Rollefson. "Axel Ragnar Olson." In *University of California: In Memoriam, 1957*, edited by University of California Academic Senate, 130–32. Berkeley: University of California, 1957. Accessed 4 July 2015. http://texts.cdlib.org/view?docId =hbow10035d.

Gleason, Sterling. "Sensational Study of Heredity May Produce New Race of Men." *Popular Science Monthly*, November 1934, 15–17.

Goldsmith, Harry. "X-Ray Produces Super-Lily Surpassing Other Blooms." *Washington Post*, 13 October 1935, B10. ProQuest.

Goodell, Rae S. "Public Involvement in the DNA Controversy: The Case of Cambridge, Massachusetts." *Science, Technology, and Human Values* 4, no. 2 (1979): 36–43.

Goodman, Jordan. "Plants, Cells, and Bodies: The Molecular Biography of Colchicine, 1930–1975." In *Molecularizing Biology and Medicine: New Practices and Alliances, 1910s–1970s*, edited by Soraya de Chadarevian and Harmke Kamminga, 17–46. Amsterdam: Harwood, 1998.

Goodspeed, T. H. "Cytological and Other Features of Variant Plants Produced from X-Rayed Sex Cells of Nicotiana Tabacum." *Botanical Gazette* 87, no. 5 (1929): 563–82.

———. "The Effects of X-Rays and Radium on Species of the Genus Nicotiana." *Journal of Heredity* 20, no. 6 (1929): 243–59.

———. "Inheritance in *Nicotiana Tabacum* IX. Mutations following Treatment with X-Rays and Radium." *University of California Publications in Botany* 11, no. 16 (1930): 285–98.

———. "Meiotic Phenomena Characteristic of First Generation Progenies from X-Rayed Tissues of *Nicotiana Tabacum*." *University of California Publications in Botany* 11, no. 18 (1930): 309–18.

———. "Occurrence of Triploid and Tetraploid Individuals in X-Ray Progenies of *Nicotiana Tabacum*." *University of California Publications in Botany* 11, no. 17 (1930): 299–308.

———. "William Albert Setchell: A Biographical Sketch." In *Essays in Geobotany: In Honor of William Albert Setchell*, edited by T. H. Goodspeed, xi–xxv. Berkeley: University of California Press, 1936.

———. "The X-Ray in Evolution." *California Monthly*, December 1927, 226–27, 256–57.

Goodspeed, T. H., and A. R. Olson. "The Production of Variation in Nicotiana Species by X-Ray Treatment of Sex Cells." *Proceedings of the National Academy of Sciences of the United States of America* 14, no. 1 (1928): 66–69.

———. "Progenies from X-Rayed Sex Cells of Tobacco." *Science* 67, no. 1724 (1928): 46.

Gottlieb, Robert. *Forcing the Spring: The Transformation of the American Environmental Movement*. Rev. ed. Washington, DC: Island Press, 2005.

Gray, George W. "New Life Made to Order." *Charleston Daily Mail Magazine*, 21 August 1932, 3, 13. NewspaperArchive.

Gregory, R. P. "On the Genetics of Tetraploid Plants in *Primula sinensis*." *Proceedings of the Royal Society of London Series B* 87, no. 597 (1914): 484–92.

Gregory, Walton C. "Induction of Useful Mutations in the Peanut." In *Brookhaven Symposia in Biology 9: Genetics in Plant Breeding*, 177–90. Upton, NY: Brookhaven National Laboratory, 1956.

———. "X-Ray Breeding of Peanuts (*Arachis Hypogaea* L.)." *Agronomy Journal* 47, no. 9 (1955): 396–99.

Griffiths, D. J., and T. D. Johnston. "The Use of an Irradiation Technique in Oat Breeding." *Radiation Botany* 2, no. 1 (1962): 41–51.

"Grow a Mystery Plant" [advertisement]. *Cleveland Plain Dealer*, 5 March 1961, E26. Plain Dealer Archive.

"Growing 'Em Bigger with Chemicals." *Popular Mechanics*, November 1940, 680–83, 136A.

"Grow Larger Grain." *Mills County Tribune*, 31 October 1912, 4. NewspaperArchive.

Grun, Paul. "The Difficulties of Defining the Term 'GM.'" *Science* 303, no. 5665 (2004): 1765–69.

Haber, Samuel. *Efficiency and Uplift: Scientific Management in the Progressive Era, 1890–1920*. Reprint, Chicago: University of Chicago Press, 1973.

Hacker, Barton C. *The Dragon's Tail: Radiation Safety in the Manhattan Project, 1942–1946*. Berkeley: University of California Press, 1987.

———. *Elements of Controversy: The Atomic Energy Commission and Radiation Safety in Nuclear Weapons Testing, 1947–1974*. Berkeley: University of California Press, 1994.

Hagen, Joel B. *An Entangled Bank: The Origins of Ecosystem Ecology.* New Brunswick, NJ: Rutgers University Press, 1992.

Hamblin, Jacob Darwin. "'A Dispassionate and Objective Effort': Negotiating the First Study on the Biological Effects of Atomic Radiation." *Journal of the History of Biology* 40, no. 1 (2007): 147–77.

———. "Let There Be Light . . . and Bread: The United Nations, the Developing World, and Atomic Energy's Green Revolution." *History and Technology* 25, no. 1 (2009): 25–48.

———. "Quickening Nature's Pulse: Atomic Agriculture at the International Atomic Energy Agency." *Dynamis* 35, no. 2 (2015): 389–408.

Hansen, Anders. "Tampering with Nature: 'Nature' and the 'Natural' in Media Coverage of Genetics and Biotechnology." *Media, Culture & Society* 28, no. 6 (2006): 811–34.

Harding, T. Swann. "Science and Agricultural Policy." In *Farmers in a Changing World: Yearbook of Agriculture, 1940,* 1081–110. Washington, DC: USDA, 1940.

Haring, Kristen. "The 'Freer Men' of Ham Radio: How a Technical Hobby Provided Social and Spatial Distance." *Technology and Culture* 44, no. 4 (2003): 734–61.

———. *Ham Radio's Technical Culture.* Cambridge, MA: MIT Press, 2007.

Harwood, Jonathan. "Peasant Friendly Plant Breeding and the Early Years of the Green Revolution in Mexico." *Agricultural History* 83, no. 3 (2009): 384–410.

Harwood, W. S. *New Creations in Plant Life: An Authoritative Account of the Life and Work of Luther Burbank.* New York: Macmillan, 1905.

———. "A Wonder-Worker of Science." *Ogden Standard,* 22 April 1905, 10. Utah Digital Newspapers.

Haskins, C. P. "X-Ray and Cathode-Ray Tubes in Biological Service. Part II." *General Electric Review* 35, no. 9 (1932): 463–71.

Haskins, C. P., and C. N. Moore. "Growth Modifications in Citrus Seedlings Grown from X-Rayed Seed." *Plant Physiology* 10, no. 1 (1935): 179–85.

———. "The Inhibition of Growth in Pollen and Mold under X-Ray and Cathode Ray Exposure." *Radiology* 23, no. 6 (1934): 710–19.

———. "X-Ray and Cathode Ray Tubes in the Service of Biology." *Radiology* 22, no. 3 (1934): 330–33.

Havas, László J. "A Colchicine Chronology." *Journal of Heredity* 31, no. 3 (1940): 115–17.

Hawkins, L. A. "Electricity and Animation." *General Electric Review* 35, no. 5 (1932): 247.

Haworth, James P. *Plant Magic.* Portland, OR: Binfords & Mort, 1946.

Hays, Samuel P. *Beauty, Health, and Permanence: Environmental Politics in the United States, 1955–1985.* Cambridge: Cambridge Univeristy Press, 1989.

Hecht, Gabrielle. *Being Nuclear: Africans and the Global Uranium Trade.* Cambridge, MA: MIT Press, 2012.

———. *The Radiance of France: Nuclear Power and National Identity after World War II.* New ed. Cambridge, MA: MIT Press, 2009.

Hedgecoe, Adam, and Paul Martin. "The Drugs Don't Work: Expectations and the Shaping of Pharmacogenetics." *Social Studies of Science* 33, no. 3 (2003): 327–64.

Henig, Robin Marantz. *Pandora's Baby: How the First Test Tube Babies Sparked the Reproductive Revolution.* New York: Houghton Mifflin, 2004.

Hesser, Leon. *The Man Who Fed the World: Nobel Peace Prize Laureate Norman Borlaug and His Battle to End World Hunger.* Dallas: Durban House, 2006.

Hewlett, Richard G., and Oscar E. Anderson Jr. *The New World, 1939/1946.* University Park: Pennsylvania State University Press, 1962.

Hewlett, Richard G., and Francis Duncan. *Atomic Shield, 1947/1952.* University Park: Pennsylvania State University Press, 1969.

Hewlett, Richard G., and Jack M. Holl. *Atoms for Peace and War, 1953–1961: Eisenhower and the Atomic Energy Commission.* Berkeley: University of California Press, 1989.

Hightower, Robin, Cathy Baden, Eva Penzes, Peter Lund, and Pamela Dunsmuir. "Expression of Antifreeze Proteins in Transgenic Plants." *Plant Molecular Biology* 17, no. 5 (1991): 1013–21.

Hillaby, John. "Atom May Unlock Arctic for Farms." *New York Times,* 16 August 1955, 6. ProQuest.

Hilts, Philip J. "The Test Tube Babies of Agriculture." *Washington Post,* 30 April 1983, A1, 8–9. ProQuest.

"Home Show Featured in Space Age Garden." *Baltimore Sun,* 5 April 1962, 16. ProQuest.

"Home Show Opens Here." *Baltimore Sun,* 10 April 1961, 11. ProQuest.

Hook, Ernest B. "James Watt Mavor (1883–1963): A Forgotten Discoverer of Radiation Effects on Heredity." *Perspectives on Biology and Medicine* 29, no. 2 (1986): 278–91.

Horlacher, W. R., and D. T. Killough. "Progressive Mutations Induced in *Gossypium Hirsutum* by Radiations." *American Naturalist* 67, no. 713 (1933): 532–38.

———. "Radiation-Induced Variation in Cotton: Somatic Changes Induced in *Gossypium Hirsutum* by X-Raying Seeds." *Journal of Heredity* 22, no. 8 (1931): 253–62.

Houghton, Kenneth W. "Experiments with Colchicine." *Horticulture,* 1 January 1940, 16.

Hounshell, David A. "The Evolution of Industrial Research in the United States." In *Engines of Innovation: U.S. Industrial Research at the End of an Era,* edited by Richard S. Rosenbloom and William J. Spencer, 13–85. Boston: Harvard Business School Press, 1996.

———. *From the American System to Mass Production, 1800–1932: The Development of Manufacturing Technology in the United States.* Baltimore: Johns Hopkins University Press, 1984.

Howorth, Muriel. *Atomic Gardening.* St. Leonards-on-Sea, UK: New World Publications, 1960.

"How to Increase World's Grains." *Marysville Evening Tribune,* 25 September 1912, 4. Newspaper-Archive.

Hughes, Sally Smith. *Genentech: The Beginnings of Biotech.* Chicago: University of Chicago Press, 2011.

———. "Making Dollars out of DNA: The First Major Patent in Biotechnology and the Commercialization of Molecular Biology, 1974–1980." *Isis* 92, no. 3 (2001): 541–75.

Hughes, Thomas P. *American Genesis: A Century of Invention and Technological Enthusiasm, 1870–1970.* Reprint, Chicago: University of Chicago Press, 2004.

———. "The Evolution of Large Technological Systems." In *The Social Construction of Technological Systems: New Directions in the Sociology and History of Technology,* anniv. ed., edited by Wiebe E. Bijker, Thomas P. Hughes, and Trevor Pinch, 45–76. Cambridge, MA: MIT Press, 2012.

———. *Networks of Power: Electrification in Western Society, 1880–1930.* Baltimore: Johns Hopkins University Press, 1983.

Hüntelmann, Axel C. "Seriality and Standardization in the Production of '606.'" *History of Science* 48, nos. 3–4 (2010): 435–60.

Imperial Bureau of Plant Genetics. *The Experimental Production of Haploids and Polyploids.* Cambridge: School of Agriculture, 1936.

"Irradiated Bulbs Add Suspense to Posy Patch." *Norfolk New Journal and Guide,* 28 April 1962, C7. ProQuest.

J.W.B. "The Lily Gets a Hypo." *Washington Post,* 28 March 1948, B8. ProQuest.

James, John. *Create New Flowers and Plants, Indoors and Out; a Gardener's Guide to Developing New Varieties through Discovery, Selection, Hybridization, and Mutation.* Garden City, NY: Doubleday, 1964.

James, Richard. "Genetic Blueprints: Horror or Hope?" *Wall Street Journal,* 11 November 1968, 18. ProQuest.

Jeffries, John W. *Wartime America: The World War II Home Front.* Chicago: I. R. Dee, 1996.

Jenkins, James A. "Roy Elwood Clausen, 1861–1956." *Biographical Memoirs of the National Academy of Sciences* 39 (1967): 37–54.

Johns, Al. "Home Show Opens at Sports Arena Thursday." *Los Angeles Times,* 21 May 1961, O1. ProQuest.

Johnson, Charles W., and Charles O. Jackson. *City Behind a Fence: Oak Ridge, Tennessee, 1942–1946.* Knoxville: University of Tennessee Press, 1981.

Johnson, Leander, and James M. Epps. "Radiation Tests Look to Wilt Resistant Cotton." *Tennessee Farm and Home Science,* July–September 1957, 4.

Johnson, Leland, and Daniel Schaffer. *Oak Ridge National Laboratory: The First Fifty Years.* Knoxville: University of Tennessee Press, 1994.

Johnson, Paige. "Safeguarding the Atom: The Nuclear Enthusiasm of Muriel Howorth." *British Journal for the History of Science* 45, no. 4 (2012): 551–71.

Jones, Jenkin W. "Improvement in Rice." In *Yearbook of Agriculture 1936,* 415–54. Washington, DC: USDA, 1936.

Kaempffert, Waldemar. "Science in the News." *New York Times,* 7 December 1941, D7. ProQuest.

Kane, Carl. "Home Show Opens with an Attendance of 2162." *Cedar Rapids Gazette,* 7 April 1961, 7. NewspaperArchive.

Kawara, Kiyoshi. "Introduction of a Gamma Field in Japan." *Radiation Botany* 3, no. 2 (1963): 175–77.

Kay, Lily E. "Life as Technology: Representing, Intervening, and Molecularizing." In *The Philosophy and History of Molecular Biology: New Perspectives,* edited by Sahotra Sarkar, 87–100. Dordrecht, Neth.: Kluwer Academic Publishers, 1996.

———. *The Molecular Vision of Life: Caltech, the Rockefeller Foundation, and the Rise of the New Biology.* New York: Oxford University Press, 1993.

Keeble, Frederick. "Giantism in *Primula Sinensis.*" *Journal of Genetics* 2, no. 2 (1912): 163–88.

Keller, Evelyn Fox. "Physics and the Emergence of Molecular Biology: A History of Cognitive and Political Synergy." *Journal of the History of Biology* 23, no. 3 (1990): 389–409.

Kelly-Campbell, Martin, and William Aspray. *Computer: A History of the Information Machine.* 2nd ed. Boulder, CO: Westview Press, 2004.

Kendall, Don. "'Hot' Seeds Produce Way out Plants." *Hutchinson News,* 2 March 1961, 2A. NewspaperArchive.

Kennerly, A. B. "New Plants on Order." *Popular Mechanics,* June 1961, 132–33, 136.

Kenney, Martin. *Biotechnology: The University-Industrial Complex.* New Haven, CT.: Yale University Press, 1986.

Kevles, Bettyann Holtzmann. *Naked to the Bone: Medical Imaging in the Twentieth Century.* Reprint, New York: Basic Books, 1998.

Kevles, Daniel J. *In the Name of Eugenics: Genetics and the Uses of Human Heredity.* Cambridge, MA.: Harvard University Press, 1995.

———. "A Primer of A, B, Seeds: Advertising, Branding, and Intellectual Property in an Emerging Industry." *UC Davis Law Review* 47, no. 2 (2013): 657–78.

Kifner, John. "'Creation of Life' Experiment at Harvard Stirs Heated Dispute." *New York Times*, 17 June 1976, 22. ProQuest.

Kimmelman, Barbara A. "The American Breeders' Association: Genetics and Eugenics in an Agricultural Context, 1903–13." *Social Studies of Science* 13, no. 2 (1983): 163–204.

———. "Mr. Blakeslee Builds His Dream House: Agricultural Institutions, Genetics, and Careers 1900–1915." *Journal of the History of Biology* 39, no. 2 (2006): 241–80.

———. "Organisms and Interests in Scientific Research: R. A. Emerson's Claims for the Unique Contributions of Agricultural Genetics." In *The Right Tools for the Job: At Work in Twentieth-Century Life Sciences*, edited by Adele E. Clarke and Joan H. Fujimura, 198–232. Princeton, NJ: Princeton University Press, 1992.

———. "A Progressive Era Discipline: Genetics at American Agricultural Colleges and Experiment Stations, 1900–1920." PhD diss., University of Pennsylvania, 1987.

Kingsbury, Noel. *Hybrid: The History and Science of Plant Breeding*. Chicago: University of Chicago Press, 2009.

Kingsland, Sharon E. "The Battling Botanist: Daniel Trembly MacDougal, Mutation Theory, and the Rise of Experimental Evolutionary Biology in America, 1900–1912." *Isis* 82, no. 3 (1991): 479–509.

Kline, Wendy. *Building a Better Race: Gender, Sexuality, and Eugenics from the Turn of the Century to the Baby Boom*. Berkeley: University of California Press, 2001.

Kloppenburg, Jack Ralph, Jr. *First the Seed: The Political Economy of Plant Biotechnology, 1492–2000*. 2nd ed. Madison: University of Wisconsin Press, 2004.

Knorr, Karin D. "Tinkering toward Success: Prelude to a Theory of Scientific Practice." *Theory and Society* 8, no. 3 (1979): 347–76.

Kohler, Robert E. *Lords of the Fly:* Drosophila *Genetics and the Experimental Life*. Chicago: University of Chicago Press, 1994.

———. "The Management of Science: The Experience of Warren Weaver and the Rockefeller Foundation Programme in Molecular Biology." *Minerva* 14, no. 3 (1976): 279–306.

———. *Partners in Science: Foundations and Natural Scientists, 1900–1945*. Chicago: University of Chicago Press, 1991.

Konzak, Calvin F. "III. Genetic Effects of Radiation on Higher Plants." *Quarterly Review of Biology* 32, no. 1 (1957): 27–45.

———. "Stem Rust Resistance in Oats Induced by Nuclear Radiation." *Agronomy Journal* 46, no. 12 (1954): 538–40.

Kopp, Carolyn. "The Origins of the American Scientific Debate over Fallout Hazards." *Social Studies of Science* 9, no. 4 (1979): 403–22.

Kostoff, Dontcho. "Chromosome Alterations by Centrifuging." *Zeitschrift für Induktive Abstammungs- und Vererbungslehre* 69, no. 1 (1935): 301–2.

Kotulak, Ronald. "And Now Your Child Built to Order." *Chicago Tribune*, 18 August 1965, 5. ProQuest.

Kraft, Ken. *Garden to Order*. Garden City, NY: Doubleday, 1963.

Krige, John. "Atoms for Peace, Scientific Internationalism, and Scientific Intelligence." *Osiris* 21 (2006): 161–81.

———. "The Politics of Phosphorus-32: A Cold War Fable Based on Fact." *Historical Studies in the Physical and Biological Sciences* 36, no. 1 (2005): 71–91.

Krimsky, Sheldon. *Biotechnics and Society: The Rise of Industrial Genetics*. New York: Praeger, 1991.

———. *Genetic Alchemy: A Social History of the Recombinant DNA Controversy*. Cambridge, MA: MIT Press, 1982.

LaFollette, Marcel C. *Making Science Our Own: Public Images of Science, 1910–1955*. Chicago: University of Chicago Press, 1990.

———. *Science on the Air: Popularizers and Personalities on Radio and Early Television*. Chicago: University of Chicago Press, 2008.

———. "Taking Science to the Marketplace: Examples of Science Service's Presentation of Chemistry during the 1930s." In *The Public Image of Chemistry*, edited by Joachim Schummer, Bernadette Bensaude-Vincent, and Brigitte Van Tiggelen, 259–96. Singapore: World Scientific, 2007.

Landecker, Hannah. *Culturing Life: How Cells Became Technologies*. Cambridge, MA: Harvard University Press, 2007.

Lang, Daniel. "Our Far-Flung Correspondents: A Stroll in the Garden." *New Yorker*, 20 July 1957, 30–59.

Larter, E. N. "A Review of the Historical Development of Triticale." In *Triticale: First Man-Made Cereal*, edited by Cho C. Tsen, 35–52. St. Paul, MN: American Association of Cereal Chemists, 1974.

Laurence, William L. "Atomic Laboratory on Long Island to Be a Mighty Research Center." *New York Times*, 1 March 1947, 17, 30. ProQuest.

———. "Science in the News." *New York Times*, 6 August 1939, D4. ProQuest.

Lavine, Matthew. *The First Atomic Age: Scientists, Radiations, and the American Public, 1895–1945*. New York: Palgrave MacMillan, 2013.

Leach, David. "Some Notes on the Induction of Mutation in Rhododendrons." *Quarterly Bulletin of the American Rhododendron Society* 4, no. 3 (1950). Accessed 15 March 2013. http://scholar.lib.vt.edu/ejournals/JARS/v4n3/v4n3-leach.htm.

Lederberg, Joshua. "Genetic Intervention Is a Way of Improving Our Species." *Washington Post and Times Herald*, 4 November 1967, A13. ProQuest.

Lenoir, Timothy, and Marguerite Hays. "The Manhattan Project for Biomedicine." In *Controlling Our Destinies: Historical, Philosophical, Ethical, and Theological Perspectives on the Human Genome Project*, edited by Phillip R. Sloan, 29–62. South Bend, IN: University of Notre Dame Press, 2000.

Lesch, John E. *The First Miracle Drugs: How the Sulfa Drugs Transformed Medicine*. Oxford: Oxford University Press, 2007.

Leung, Victoria. "Between Farming and Radioscience: A Study of Induced Mutation Breeding, c1920–1980." Master's thesis, Imperial College London, 2007.

Libby, Willard F. "The Economic Potential of Radioisotopes in Agriculture." In *A Conference on Radioactive Isotopes in Agriculture; held on January 12, 13 and 14, 1956 at Michigan State University, East Lansing, Michigan*, 1–6. Washington, DC: United States Atomic Energy Commission, 1956.

Lindee, M. Susan. *Suffering Made Real: American Science and the Survivors at Hiroshima*. Chicago: University of Chicago Press, 1994.

Lindsay, Christina. "From the Shadows: Users as Designers, Producers, Marketers, Distributors, and Technical Support." In *How Users Matter: The Co-construction of Users and Technology*, edited by Nelly Oudshoorn and Trevor Pinch, 29–50. Cambridge, MA: MIT Press, 2003.

Lipartito, Kenneth. "Picturephone and the Information Age: The Social Meaning of Failure." *Technology and Culture* 44, no. 1 (2003): 50–81.

Liu, L. X., H. J. Guo, L. S. Zhao, J. Wang, J. Y. Gu, and S. R. Zhao. "Achievements and Perspectives of Crop Space Breeding in China." In *Induced Plant Mutations in the Genomics Era*, edited by Q. Y. Shu, 213–15. Rome: FAO, 2009.

Logan, H. Britton, Jean-Marie Putnam, and Lloyd C. Cosper. *Science in the Garden*. New York: Duell, Sloan and Pearce, 1941.

Lundqvist, U. "Eighty Years of Scandinavian Barley Mutation Genetics and Breeding." In *Induced Plant Mutations in the Genomics Era*, edited by Q. Y. Shu, 39–43. Rome: FAO, 2009.

Lurquin, Paul F. *The Green Phoenix: A History of Genetically Modified Plants*. New York: Columbia University Press, 2001.

Lutz, Anne M. "A Preliminary Note on the Chromosomes of Oenothera Lamarckiana and One of Its Mutants, O. Gigas." *Science* 26, no. 657 (1907): 151–52.

Lyon-Jenness, Cheryl. "Planting a Seed: The Nineteenth-Century Horticultural Boom in America." *Business History Review* 78, no. 3 (2004): 381–421.

MacKenzie, Donald, and Judy Wajcman, eds. *The Social Shaping of Technology*. 2nd ed. Buckingham, UK: Open University Press, 1999.

MacKey, James. "Mutation Breeding in Europe." In *Brookhaven Symposia in Biology 9: Genetics in Plant Breeding*, 141–56. Upton, NY: Brookhaven National Laboratory, 1956.

"Magic-Ray Farmers." *Popular Mechanics*, July 1933, 58–61, 120A, 24A.

Manchester, Harland. "The New Age of 'Atomic Crops.'" *Popular Mechanics*, October 1958, 106–10, 282, 284, 286, 288.

Mandeville, Edna King. "'Atomic' Seeds Offer Experiment Chance." *Hutchinson News*, 30 April 1961, 23. NewspaperArchive.

Mangelsdorf, Paul C. *Corn: Its Origin, Evolution, and Improvement*. Cambridge, MA: Harvard University Press, 1974.

Marchand, Roland. *Advertising the American Dream: Making Way for Modernity, 1920–1940*. Berkeley: University of California Press, 1985.

Martin, John H. "Sorghum Improvement." In *Yearbook of Agriculture 1936*, 523–60. Washington, DC: USDA, 1936.

Masco, Joseph. "Mutant Ecologies: Radioactive Life in Post-Cold War New Mexico." *Cultural Anthropology* 19, no. 4 (2004): 517–50.

Mavor, James W. "The Attack on the Gene." *Scientific Monthly* 21, no. 4 (1925): 355–63.

———. "An Effect of X Rays on the Linkage of Mendelian Characters in the First Chromosome of Drosophila." *Genetics* 8, no. 4 (1923): 355–66.

———. "On the Elimination of the X-Chromosome from the Egg of Drosophila Melanogaster by X-Rays." *Science* 54, no. 1395 (1921): 277–79.

———. "The Production of Non-Disjunction by X-Rays." *Science* 55, no. 1420 (1922): 295–97.

———. "The Production of Non-Disjunction by X-Rays." *Journal of Experimental Zoology* 39, no. 2 (1924): 381–432.

Mavor, James W., and Henry K. Svenson. "X-Rays and Crossingover." *Science* 58, no. 1494 (1923): 124–26.

May, Lee. "Battle over Plant Gene Splicing Heats Up." *Los Angeles Times*, 3 November 1983, D1. ProQuest.

Mazuzan, George T., and J. Samuel Walker. *Controlling the Atom: The Beginnings of Nuclear Regulation, 1946–1962*. Berkeley: University of California Press, 1985.

McAfee, Kathleen. "Biotech Battles: Plants, Power, and Intellectual Property in the New Global Governance Regimes." In *Engineering Trouble: Biotechnology and Its Discontents*, edited by Rachel A. Schurman and Dennis Doyle Takahashi Kelso, 174–94. Berkeley: University of California Press, 2003.

McCaig, Imogene. "Space Seedsman." *Christian Science Monitor*, 18 July 1961, 8. ProQuest.

McDonough, Stephen J. "Agriculture Dept. Developes [sic] New Species of Plants." *Hartford Courant*, 20 August 1939, SM11. ProQuest.

McKay, J. W., and T. H. Goodspeed. "The Effects of X-Radiation on Cotton." *Science* 71, no. 1851 (1930): 644.

McManus, Robert Cruise. "Agriculture's Giant Powder." *Country Home Magazine*, April 1939, 13, 29–30.

Medhurst, Martin J. "Atoms for Peace and Nuclear Hegemony: The Rhetorical Structure of a Cold War Campaign." *Armed Forces & Society* 23, no. 4 (1997): 571–93.

"Meet Dr. C. J. Speas" [advertisement]. *Cleveland Plain Dealer*, 5 March 1961, F10. Plain Dealer Archive.

Meikle, Jeffrey L. *American Plastic: A Cultural History*. New Brunswick, NJ: Rutgers University Press, 1995.

"Men and Things: Speeding up Evolution." *American Medicine* 25, no. 8 (1919): 512.

Meyer-Thurow, George. "The Industrialization of Invention: A Case Study from the German Chemical Industry." *Isis* 73, no. 3 (1982): 363–81.

Moh, C. C., R. A. Nilan, and F. C. Elliott. "An Unusual Association of Two Mutant Characters in Atom-Bombed Barley." *Journal of Heredity* 46, no. 1 (1955): 35–40.

Moh, C. C., and G. Orbegoso. "Effects of Ionizing Radiations on Coffee." *Coffee and Cacao Technical Services* 1, no. 2 (1959): 25–30.

Moore, C. N., and C. P. Haskins. "X-Ray Induced Modifications of Flower Color in the Petunia." *Journal of Heredity* 26, no. 9 (1935): 349–55.

Moore, Kelly. *Disrupting Science: Social Movements, American Scientists, and the Politics of the Military, 1945–1975*. Princeton, NJ: Princeton University Press, 2008.

Morange, Michel. *A History of Molecular Biology*. Translated by Matthew Cobb. Cambridge, MA: Harvard University Press, 1998.

"More Food with Atom-Blasted Seeds." *Times of India*, 6 September 1959, 7. ProQuest.

Morgan, Thomas Hunt. *The Physical Basis of Heredity*. Philadelphia: J. B. Lippincott, 1919.

Morgan, T. H., C. B. Bridges, and A. H. Sturtevant. "The Genetics of Drosophila." *Bibliographia Genetica* 2 (1925): 1–262.

Morrison, A. Cressy. *Man in a Chemical World: The Service of Chemical Industry*. New York: C. Scribner's Sons, 1937.

Morrison, Gordon. "Facts about Colchicine." *Gardeners' Chronicle of America*, October 1939, 297, 326.

Mowery, David C., and Nathan Rosenberg. *Paths of Innovation: Technological Change in 20th Century America*. Cambridge: Cambridge University Press, 1998.

Muller, H. J. "Artificial Transmutation of the Gene." *Science* 66, no. 1699 (1927): 84–87.

———. "The Method of Evolution." *Scientific Monthly* 29, no. 6 (1929): 481–505.

———. "Why Polyploidy Is Rarer in Animals than in Plants." *American Naturalist* 59, no. 663 (1925): 346–53.

Müntzing, Arne. "The Evolutionary Significance of Autopolyploidy." *Hereditas* 21, no. 2–3 (1936): 363–78.

Murphy, Denis. *Plant Breeding and Biotechnology: Societal Context and the Future of Agriculture*. Cambridge: Cambridge University Press, 2007.

"Musician Speeds Up Tree Growth in X-Ray Experiments." *Los Angeles Times*, 23 February 1937, 16. ProQuest.

"National Food Stores, 'Atomic Gardening'" [advertisement]. *Chicago Daily Tribune*, 23 April 1961, W–B18. ProQuest.

National Research Council. *Triticale: A Promising Addition to the World's Cereal Grains*. Reprint, New York: Books for Business, 2002.

Nebel, Bernard R. "Cytological Observations of Colchicine." *Biological Bulletin* 73, no. 2 (1937): 351–52.

———. "Inducing Changes in Plants with Colchicine Shows Progress." *Farm Research* 6, no. 1 (1940): 10, 15.

Nebel, B. R., and M. L. Ruttle. "Colchicine and Its Place in Fruit Breeding." *Circular of the New York State Agricultural Experimental Station* 183 (1938): 1–19.

———. "The Cytological and Genetical Significance of Colchicine." *Journal of Heredity* 29, no. 1 (1938): 3–9.

Nelkin, Dorothy, and M. Susan Lindee. *The DNA Mystique*. 2nd ed. Ann Arbor: University of Michigan Press, 2004.

"New Knowledge Hot Corn." *Pathfinder*, 19 September 1951.

Newman, Iva. "New Aids for Garden Shown at Exposition." *San Mateo Times*, 2 February 1962, 9. NewspaperArchive.

"New Plant Varieties, Improved Animal Breeds Expected from X-Ray." *Lawrence Journal-World*, 24 July 1930, 6. Google News Archive.

Nims, L. F. "Opportunities for Research in Biology." In *Brookhaven Conference Report: Biology and Medicine, 16–18 October 1947*, 3–4. Upton, NY: Brookhaven National Laboratory, 1947.

Nisbet, Matthew C., and Bruce V. Lewenstein. "Biotechnology and the American Media: The Policy Process and the Elite Press, 1970–1999." *Science Communication* 23, no. 4 (2002): 359–91.

"Nuclear Age Means More Food for Mankind." *Times of India*, 12 August 1955, 8. ProQuest.

Nutch, Frank. "Gadgets, Gizmos, and Instruments: Science for the Tinkering." *Science, Technology, and Human Values* 21, no. 2 (1996): 214–28.

Nye, David E. *Consuming Power: A Social History of American Energies*. Cambridge, MA: MIT Press, 1998.

———. *Electrifying America: Social Meanings of a New Technology, 1880–1940*. Cambridge, MA: MIT Press, 1990.

Nye, Patrick W. "An Oral History of Haskins Laboratory, December 2006" [interview with Caryl Haskins, Franklin Cooper, and Seymour Hutner]. Haskins Laboratories, 2007. Accessed 2 July 2015. http://www.haskins.yale.edu/history/oh/HL_Oral_History.pdf.

Oatsvall, Neil. "Atomic Agriculture: Policymaking, Food Production, and Nuclear Technologies in the United States, 1945–1960." *Agricultural History* 88, no. 3 (2014): 368–87.

Olby, Robert C. *Origins of Mendelism*. 2nd ed. Chicago: University of Chicago Press, 1985.

"Old Gout Remedy May Mean Revolution on the Farm." *Spokane Daily Chronicle*, 29 December 1937, 9. Google News Archive.

Oliver, Warner. "Horizons in Test Tubes." *Lincoln Sunday Journal and Star*, 12 November 1939, E4.

Olmstead, Alan L., and Paul W. Rhode. *Creating Abundance: Biological Innovation and American Agricultural Development*. Cambridge: Cambridge University Press, 2008.

Olson, A. R. "The Effect of X-Rays on Chemical Reactions." *Science* 56, no. 1443 (1922): 231.

Olson, A. R., Elmer Dershem, and H. H. Storch. "X-Ray Absorption Coefficients of Carbon, Hydrogen and Oxygen." *Physical Review* 21, no. 1 (1923): 30.

O'Neil, Paul. "Yaphank's Happy Thinkers," *Life*, 29 September 1958, 102–18.

Orr, Richard. "The Home Garden." *Chicago Daily Tribune*, 13 April 1961.

Osborne, T. S. "Atomic Tools Help Plant Breeders." *Tennessee Farm and Home Science*, April–June 1956, 3, 9.

————. "Radiation and Plant Breeding." *Tennessee Farm and Home Science*, April–June 1957, 8.

————. "Regional and National Programs on Use of Irradiation in Plant Breeding." In *Proceedings of Southeastern Seminar on Atomic Progress in Agriculture, Clemson House, February 13, 14, 1961*, 36–43. Clemson, SC: Clemson College, 1961.

Osborne, T. S., and A. O. Lunden. "The Cooperative Plant and Seed Irradiation Program of the University of Tennessee." *International Journal of Applied Radiation and Isotopes* 10, no. 4 (1961): 198–209.

Oudshoorn, Nelly, and Trevor Pinch, eds., *How Users Matter: The Co-construction of Users and Technology*. Cambridge, MA: MIT Press, 2003.

Palladino, Paolo. "Wizards and Devotees: On the Mendelian Theory of Inheritance and the Professionalization of Agricultural Science in Great Britain and the United States, 1880–1930." *History of Science* 32, no. 4 (1994): 409–44.

Palumbi, Stephen R. "Humans as the World's Greatest Evolutionary Force." *Science* 293, no. 5536 (2001): 1786–90.

Pandora, Katherine. "Knowledge Held in Common: Tales of Luther Burbank and Science in the American Vernacular." *Isis* 92, no. 3 (2001): 484–516.

"Patents of the Week." *Science News-Letter* 84, no. 15 (1963): 238.

Paul, Diane B. *Controlling Human Heredity: 1865 to the Present*. Atlantic Highlands, NJ: Humanities Press International, 1995.

Paul, Diane B., and Barbara A. Kimmelman. "Mendel in America: Theory and Practice, 1900–1919." In *The American Development of Biology*, edited by Ronald Rainger, Keith R. Benson, and Jane Maienschein, 281–310. Philadelphia: University of Pennsylvania Press, 1988.

Pauly, Philip J. *Controlling Life: Jacques Loeb and the Engineering Ideal in Biology*. New York: Oxford University Press, 1987.

————. *Fruits and Plains: The Horticultural Transformation of America*. Cambridge, MA: Harvard University Press, 2007.

"Pepped-up Peanut Plant Will Be Shown Tomorrow." *Cleveland Plain Dealer*, 31 July 1960, F14. Plain Dealer Archive.

Perkins, John H. *Geopolitics and the Green Revolution: Wheat, Genes, and the Cold War*. New York: Oxford University Press, 1997.

"Planters Now Can Predict How Well Seeds Will Grow." *Kingsport News*, 1 December 1958, 2. NewspaperArchive.

"Plant Evolution Can Be Effected by X-Ray." *Chillicothe Constitution-Tribune*, 2 April 1928, 5. NewspaperArchive.

"Plant Magicians Now at Work Improving World's Food Crops." *Newsweek*, May 1944, 94, 96–99.

"Plants Doped to Make Them Produce More." *Frederick News Post*, 23 September 1939, 8. NewspaperArchive.

"Plants Grown to Order." *Popular Mechanics*, April 1943, 56–58, 156.

Plough, Harold H. "The Effect of Temperature on Crossingover in Drosophila." *Journal of Experimental Zoology* 24, no. 2 (1917): 147–209.

Poehlman, John Milton. *A History of Field Crops 1870–1967 in the University of Missouri*. Agricultural Experiment Station Special Report 385. College of Agriculture, University of Missouri–Columbia, December 1988.

Pollock, James B. "Some Physiological Variations of Plants, and Their General Significance." *Science* 25, no. 649 (1907): 881–89.

Porter, Roy. "The Prince's Poison." *New Scientist* 169, no. 2274 (2001): 42–43.

Potrykus, Ingo. "Golden Rice and Beyond." *Plant Physiology* 125, no. 3 (2001): 1157–61.

Pottage, Alain, and Brad Sherman. "Organisms and Manufactures: On the History of Plant Inventions." *Melbourne University Law Review* 31, no. 2 (2007): 539–68.

Prakash, Channapatna S. "The Genetically Modified Crop Debate in the Context of Agricultural Evolution." *Plant Physiology* 126, no. 1 (2001): 8–15.

Prather, Robert H. "Better Nutrition." *Sierra Educational News*, May 1943, 2–3.

Pringle, Peter. *Food, Inc.: Mendel to Monsanto—the Promises and Perils of the Biotech Harvest.* New York: Simon & Schuster, 2003.

Proceedings of the International Conference on the Peaceful Uses of Atomic Energy. Vol. 12, *Radioactive Isotopes and Ionizing Radiations in Agriculture, Physiology, and Biochemistry.* New York: United Nations, 1956.

"Purple Tomato." *Hartford Courant*, 27 May 1961, 10. ProQuest.

Pursell, Carroll, ed. *A Companion to American Technology.* Oxford: Blackwell, 2008.

———. "The Farm Chemurgic Council and the United States Department of Agriculture, 1935–1939." *Isis* 60, no. 3 (1969): 307–17.

———. *The Machine in America: A Social History of American Technology.* 2nd ed. Baltimore: Johns Hopkins University Press, 2007.

R.W. Review of *Cytology of Fruits as Related to Plant Breeding*, by B. R. Nebel. *Agronomy Journal* 29, no. 4 (1937): 335.

Rader, Karen. "Alexander Hollaender's Postwar Vision for Biology: Oak Ridge and Beyond." *Journal of the History of Biology* 39, no. 4 (2006): 685–706.

"Radiation and the Green Revolution." *International Atomic Energy Agency Bulletin* 11, no. 5 (1969): 16–27.

"Radiation Used to 'Speed Up' Evolution, Scientists Disclose." *Newsday*, 2 April 1954, 5S.

Randolph, L. F. "An Evaluation of Induced Polyploidy as a Method of Breeding Crop Plants." *American Naturalist* 75, no. 759 (1941): 347–63.

———. "Some Effects of High Temperature on Polyploidy and Other Variations in Maize." *Proceedings of the National Academy of Sciences of the United States of America* 18, no. 3 (1932): 222–29.

Rasmussen, Nicolas. "The Forgotten Promise of Thiamin: Merck, Caltech Biologists, and Plant Hormones in a 1930s Biotechnology Project." *Journal of the History of Biology* 32, no. 2 (1999): 245–61.

———. *Gene Jockeys: Life Science and the Rise of Biotech Enterprise.* Baltimore: Johns Hopkins University Press, 2014.

———. "Making a Machine Instrumental: RCA and the Wartime Origins of Biological Electron Microscopy in America, 1940–1945." *Studies in History and Philosophy of Science Part A* 27, no. 3 (1996): 311–49.

———. "Plant Hormones in War and Peace: Science, Industry, and Government in the Development of Herbicides in 1940s America." *Isis* 92, no. 2 (2001): 291–316.

Ray, G. E. "Plant Magic Is World Hobby." *Flower Grower*, 1942, 175–76.

Rédei, G. P. "A Portrait of Lewis John Stadler, 1896–1954." In *Stadler Genetics Symposia*, vol. 1/2, 5–20. Columbia: University of Missouri–Columbia, 1971.

Reed, H. S., and T. S. Osborne. "Soybean Research in Tennessee." *Soybean Digest* 19, no. 5 (1959): 18–19.

Reich, Leonard S. *The Making of American Industrial Research: Science and Business at GE and Bell, 1876–1926.* Cambridge: Cambridge University Press, 1985.

Reinhold, Robert. "Evolution Control: A Genetic Advance." *New York Times*, 8 September 1968, 1. ProQuest.

Rentetzi, Maria. "Rackaging Radium, Selling Science: Boxes, Bottles and Other Mundane Things in the World of Science." *Annals of Science* 68, no. 3 (2011): 375–99.

"Reported from the Field of Science." *New York Times*, 6 October 1940, 59. ProQuest.

"Reports of the Sessions of Sections and Societies at the Second Nashville Meeting." *Science* 67, no. 1727 (1928): 113–31.

"Researchers Create the World's First Fully Synthetic, Self-Replicating Living Cell." *PopSci*, 20 May 2010. Accessed 17 December 2011. http://www.popsci.com/science/article/2010-05/j-craig-venter-institute-creates-worlds-first-synthetic-cell.

Rheinberger, Hans-Jörg, and Jean-Paul Gaudillière, eds. *Classical Genetic Research and Its Legacy: The Mapping Cultures of Twentieth-Century Genetics*. London: Routledge, 2004.

Rhoades, M. M. "The Early Years of Maize Genetics." *Annual Review of Genetics* 18 (1984): 1–29.

———. "Lewis John Stadler, 1896–1954." *Biographical Memoirs of the National Academy of Sciences* 30 (1957): 328–47.

Rhodes, Richard. *The Making of the Atomic Bomb*. New York: Simon & Schuster, 1986.

Richmond, Marsha L. "Women in Mutation Studies: The Role of Gender in the Methods, Practices, and Results of Early Twentieth-Century Genetics." In "Making Mutations: Objects, Practices, Contexts," edited by Luis Campos and Alexander von Schwerin, 11–48. Preprint 393. Berlin: Max Planck Institute for the History of Science, 2010.

Richter, Alan, and W. Ralph Singleton. "The Effect of Chronic Gamma Radiation on the Production of Somatic Mutations in Carnations." *Proceedings of the National Academy of Sciences of the United States of America* 41, no. 5 (1955): 295–300.

Riley, Morgan T. *Dahlias: What Is Known about Them*. New York: Orange Judd, 1947.

Roberts, J. A. Fraser. "Reginald Ruggles Gates. 1882–1962." *Biographical Memoirs of Fellows of the Royal Society* 10 (1964): 83–106.

Rockwell, F. F. "'Round about the Garden." *New York Times*, 24 July 1938, 38. ProQuest.

Roessner, Elmer. "Atomic Age Hits the Garden as Irradiated Plants Go on the Market." *Boston Globe*, 15 April 1962, B12. ProQuest.

Roman, Herschel. "A Diamond in a Desert." *Genetics* 119, no. 4 (1988): 739–41.

Romani, R. J., E. C. Maxie, C. O. Hesse, and N. F. Sommer. "Cobalt-60 Gamma-Ray Irradiator Opens New Doors to Biological Research at Davis." *California Agriculture*, March 1962, 2–4.

Rosenberg, Charles E. *No Other Gods: On Science and American Social Thought*. Rev. and expanded ed. Baltimore: Johns Hopkins University Press, 1997.

Russell, Edmund. *War and Nature: Fighting Humans and Insects with Chemicals from World War I to Silent Spring*. Cambridge: Cambridge University Press, 2001.

Ruttle, M. L. "Colchicine and the Production of Polyploid Essential Oil Plants." *Herbarist* 5 (1939): 20–27.

Ruttle, M. L., and B. R. Nebel. "Extend Use of Colchicine as Plant Breeding Tool." *Farm Research* 7, no. 1 (1941): 10, 15.

Sagawa, Yoneo, and Gustav A. L. Mehlquist. "The Mechanism Responsible for Some X-Ray Induced Changes in Flower Color of the Carnation, Dianthus Caryophyllus." *American Journal of Botany* 44, no. 5 (1957): 397–403.

Sale, Kirkpatrick. *The Green Revolution: The American Environmental Movement, 1962–1992*. New York: Hill and Wang, 1993.

Santesmases, María Jesús. "Cereals, Chromosomes and Colchicine: Crop Varieties at the Estación Experimental Aula Dei and Human Cytogenetics, 1948–58." In *Human Heredity in the Twentieth Century*, edited by Bernd Gausemeier, Staffan Müller-Wille, and Edmund Ramsden, 127–40. London: Pickering & Chatto, 2013.

———. "Peace Propaganda and Biomedical Experimentation: Influential Uses of Radioisotopes in Endocrinology and Molecular Genetics in Spain (1947–1971)." *Journal of the History of Biology* 39, no. 4 (2006): 765–94.

Sapp, Jan. "The Nine Lives of Gregor Mendel." In *Experimental Inquiries: Historical, Philosophical and Social Studies of Experimentation in Science*, edited by H. E. Le Grand, 137–66. Dordrecht, Neth.: Kluwer, 1990.

Sax, Karl. "Effect of Variations in Temperature on Nuclear and Cell Division in Tradescantia." *American Journal of Botany* 24, no. 4 (1937): 218–25.

———. "The Experimental Production of Polyploidy." *Journal of the Arnold Arboretum* 17 (1936): 153–59.

———. "The Relation between Chromosome Number, Morphological Characters and Rust Resistance in Segregates of Partially Sterile Wheat Hybrids." *Genetics* 8, no. 4 (1923): 301–21.

"Says X-Rayed Eggs Hatch Mostly Hens." *New York Times*, 1 May 1928, 25. ProQuest.

Schaffer, Simon. "Easily Cracked: Scientific Instruments in States of Disrepair." *Isis* 102, no. 4 (2011): 706–17.

Schrepfer, Susan R., and Philip Scranton, eds. *Industrializing Organisms: Introducing Evolutionary History*. New York: Routledge, 2004.

Schurman, Rachel, and William A. Munro. *Fighting for the Future of Food: Activists versus Agribusiness in the Struggle over Biotechnology*. Minneapolis: University of Minnesota Press, 2010.

Schwartz, James. *In Pursuit of the Gene: From Darwin to DNA*. Cambridge, MA: Harvard University Press, 2008.

"Science: GE's Lily." *Time*, 9 September 1935. Accessed 24 February 2012. http://www.time.com/time/magazine/article/0,9171,748972,00.html.

"Science: New Flowers by X-Rays." *Time*, 2 February 1942. Accessed 24 February 2012. http://www.time.com/time/magazine/article/0,9171,849785,00.html.

"Science Opens the Way to Make Our Children Giants." *American Weekly*, 21 November 1937, 7.

"Scientist Converts Tall Field Corn into Short for Easier Harvesting." *New York Times*, 27 August 1948, 1. ProQuest.

"Scientists Grow Peculiar Tobacco Plants with X-Ray." *Charleston Gazette*, 30 October 1927, Part 2, 7. NewspaperArchive.

"Scientists Keep an Eye on 'Atomic Seeds.'" *Cleveland Plain Dealer*, 15 January 1961, G5. Plain Dealer Archive.

Scranton, Philip. *Endless Novelty: Specialty Production and American Industrialization, 1865–1925*. Princeton, NJ: Princeton University Press, 1997.

Sears, E. R. "Amphidiploids in the *Triticinae* Induced by Colchicine." *Journal of Heredity* 30, no. 2 (1939): 38–43.

"Secret of Life Sought in Mystery Ray." *Popular Mechanics*, October 1930, 564–67.

"Seeks Cross That Will Stay Crossed." *Kingsport News*, 17 July 1961, 2. NewspaperArchive.

Serviss, Garrett P. "Transforming the World of Plants." *Cosmopolitan*, November 1905, 63–70.

Severson, Kim. "Watermelons Get Small." *New York Times*, 17 August 2010, D1. Accessed 11 August 2015. http://www.nytimes.com/2010/08/18/dining/18melons.html.

Shabecoff, Philip. *Fierce Green Fire: The American Environmental Movement.* Revised ed. Washington, DC: Island Press, 2003.

Shapiro, Seymour. "The Brookhaven Radiations Mutation Program." In *A Conference on Radioactive Isotopes in Agriculture; held on January 12, 13 and 14, 1956 at Michigan State University, East Lansing, Michigan,* 141–50. Washington, DC: United States Atomic Energy Commission, 1956.

Shaw, D. John. *World Food Security: A History Since 1945.* Basingstoke, UK: Palgrave Macmillan, 2007.

Shearer, Lloyd. "How Soon Can We Change the Human Race?" *Boston Globe,* 4 January 1970, D6.

"'Shots in Arm' for Plants Are Used to Improve Breed." *Racine Journal-Times,* 21 September 1939, sec. 2, 10. NewspaperArchive.

Shull, George Harrison. "Importance of the Mutation Theory in Practical Breeding." *Journal of Heredity* 3, no. 1 (1907): 60–67.

Shulman, Bruce J. *The Seventies: The Great Shift in American Culture, Society, and Politics.* New York: Free Press, 2001.

Silow, R. A. "The Potential Contribution of Atomic Energy to Development in Agriculture and Related Industries." *International Journal of Applied Radiation and Isotopes* 3, no. 4 (1958): 257–80.

Singleton, W. Ralph. "Atomic Energy and Agriculture." *Virginia Economic Review* 10, no. 2 (1957): 1–6.

———. "Irradiated Seed: An Expert Measures the Prospects." *New York Times,* 13 May 1962, X23. ProQuest.

———, ed. *Nuclear Radiation in Food and Agriculture.* Geneva Series on the Peaceful Uses of Atomic Energy. Princeton, NJ: Van Nostrand, 1958.

———. "The Use of Radiation in Plant Breeding. Part II." *Seed World,* February 1960, 6–7.

Sinnott, Edmund W. "Albert Francis Blakeslee, 1874–1954." *Biographical Memoirs of the National Academy of Sciences* 33 (1959): 1–38.

Sinnott, Edmund W., and L. C. Dunn. *Principles of Genetics: An Elementary Text, with Problems.* New York: McGraw-Hill, 1925.

Slater, Leo B. "Malaria Chemotherapy and the 'Kaleidoscopic' Organization of Biomedical Research during World War II." *Ambix* 51, no. 2 (2004): 107–34.

Smith, Alice Kimball. *A Peril and a Hope: The Scientists' Movement in America, 1945–47.* Chicago: University of Chicago Press, 1965.

Smith, Harold H. "Radiation in the Production of Useful Mutations." *Botanical Review* 24, no. 1 (1958): 1–24.

Smith, Jane S. *The Garden of Invention: Luther Burbank and the Business of Breeding Plants.* New York: Penguin, 2009.

Smith, John Kenly, Jr. "The Scientific Tradition in American Industrial Research." *Technology and Culture* 31, no. 1 (1990): 121–31.

Smith, Luther. "A Rare Dominant Chlorophyll Mutant in Durum Wheat Induced by Atomic Bomb Irradiation." *Journal of Heredity* 43, no. 3 (1952): 125–28.

Smith, Philip H. "No Short-Cut Horticulture." *Scientific American,* September 1940, 140–42.

Smocovitis, Vassiliki Betty. "Botany and the Evolutionary Synthesis: The Life and Work of G. Ledyard Stebbins, Jr." PhD diss., Cornell University, 1988.

———. "The 'Plant *Drosophila*': E. B. Babcock, the Genus *Crepis*, and the Evolution of a Genetics Research Program at Berkeley, 1915–1947." *Historical Studies in the Natural Sciences* 39, no. 3 (2009): 300–55.

Sonnedecker, Glenn. "Fuel for Fighters." *Science News-Letter,* 5 September 1942, 154–56.

Sparrow, Arnold H., and W. Ralph Singleton. "The Use of Radiocobalt as a Source of Gamma Rays and Some Effects of Chronic Irradiation on Growing Plants." *American Naturalist* 87, no. 832 (1953): 29–48.

"The Speas Story: It All Started on a Hillside." *Cleveland Plain Dealer*, 13 November 1960, G1. Plain Dealer Archive.

"Species Improved, Sex Is Determined by X-Ray on Eggs." *Washington Post*, 1 May 1928, 8. ProQuest.

"Speeds Breeding Types." *Los Angeles Times*, 2 October 1927, 6. ProQuest.

Springer, Anthony M. "Early Experimental Programs of the American Rocket Society, 1931–1940." *Journal of Spacecraft and Rockets* 40, no. 4 (2003): 475–90.

"The Spyglass." *Belleville Telescope*, 3 May 1962, B-1. NewspaperArchive.

Stadler, L. J. "Genetic Effects of X-Rays in Maize." *Proceedings of the National Academy of Sciences of the United States of America* 14, no. 1 (1928): 69–75.

———. "Mutations in Barley Induced by X-Rays and Radium." *Science* 68, no. 1756 (1928): 186–87.

———. "Some Genetic Effects of X-Rays in Plants." *Journal of Heredity* 21, no. 1 (1930): 3–20.

———. "The Variability of Crossing over in Maize." *Genetics* 11, no. 1 (1926): 1–37.

———. "Variation in the Intensity of Linkage in Maize." *American Naturalist* 59, no. 663 (1925): 366–70.

Stamhuis, Ida H., Onno G. Meijer, and Erik J. A. Zevenhuizen. "Hugo de Vries on Heredity, 1889–1903: Statistics, Mendelian Laws, Pangenes, Mutations." *Isis* 90, no. 2 (1999): 238–67.

Stephenson, William. "New Plant World A'Coming!" *Coronet*, December 1945, 136–39.

Stern, Alexandra Minna. *Eugenic Nation: Faults and Frontiers of Better Breeding in Modern America.* Berkeley: University of California Press, 2005.

Stokley, James. "Sir Oliver Lodge Expounds His Cosmogony." *Science News-Letter* 12, no. 336 (1927): 185–86.

Stroman, Govan N., and Thomas H. Lewis. "A Study of Genetic Effects of Cosmic Radiation on Cotton Seed." *Journal of Heredity* 42, no. 4 (1951): 211–13.

"Students Check 'Atomic Seeds.'" *Cleveland Plain Dealer*, 8 January 1961, F15. Plain Dealer Archive.

Sturtevant, A. H. *A History of Genetics.* New York: Harper & Row, 1965.

Swingle, Walter T. "New Citrous Fruits." *Journal of Heredity* 4, no. 2 (1913): 83–95.

"A Symposium of Some Activities of the Research Laboratory of the General Electric Company." In *Transactions of the American Institute of Chemical Engineers.* Vol. 28, *1932*, 31–55. New York: Van Nostrand, 1933.

Takahashi, Yuzo. "A Network of Tinkerers: The Advent of the Radio and Television Receiver Industry in Japan." *Technology and Culture* 41, no. 3 (2000): 460–84.

Taloumis, George. "Atomic Gardening Produces Remarkable Fruits and Vegetables." *Boston Globe*, 16 April 1961, 20B. ProQuest.

Tauger, Mark B. *Agriculture in World History.* London: Routledge, 2011.

Taylor, Frank J. "How Your Flowers Are Remodeled." *Saturday Evening Post*, 1 March 1947, 20, 54–56.

———. "Nature Gets the Speed Up." *Better Homes and Gardens*, May 1940, 100–102.

———. "Speeding up Nature." *Reader's Digest*, May 1940, 69–71.

———. "They've Grown Some New Posies for the Ladies." *Saturday Evening Post*, 17 March 1951, 21, 134–35, 137.

Teale, Edwin. "Test-Tube Magic Creates Amazing New Flowers." *Popular Science Monthly*, May 1940, 57–64, 227.

Teas, Howard J. "Station Installs Cobalt Irradiator." *Sunshine State Agricultural Research Report* 3 (1958): 4–5.

———. "The Use of Cobalt-60 Gamma Radiation in Ornamental Horticulture." *Florida State Horticultural Society* 71 (1958): 450–52.

"Tells Effect of Radium on Plant Reproductions." *Berkeley Daily Gazette*, 19 February 1929, 3. Google News Archive.

Thackray, Arnold, ed. *Private Science: Biotechnology and the Rise of the Molecular Sciences*. Philadelphia: University of Pennsylvania Press, 1998.

Theunissen, Bert. "Knowledge Is Power: Hugo de Vries on Science, Heredity and Social Progress." *British Journal for the History of Science* 27, no. 3 (1994): 291–311.

Thomson, Betty F. "New Kinds of Plants by Chemical Treatment." *Plants and Gardens*, Summer 1948, 117–23.

Thone, Frank. "Amateur Plant Breeders Aid Science." *Bradford Era*, 17 December 1941, 10. NewspaperArchive.

———. "New Plants and Animals Developed by Magic X-Ray." *Corpus Christi Times*, 20 June 1930, 8. NewspaperArchive.

———. "Science Prize Winner Makes Discovery Believed by Many to Have Immense Practical and Scientific Significance." *Scientific American*, March 1928, 235.

———. "Science Stunts for the Gardener." *Science News-Letter*, 13 April 1940, 234–36.

———. "X-Rays Speed Up Evolution over 1,000 Per Cent." *Science News-Letter* 12, no. 340 (1927): 243–46.

Tinn, Honghong. "From DIY Computers to Illegal Copies: The Controversy over Tinkering with Microcomputers in Taiwan, 1980–1984." *IEEE Annals of the History of Computing* 33, no. 2 (2011): 75–88.

Titus, A. Constandina. "Selling the Bomb: Public Relations Efforts by the Atomic Energy Commission during the 1950s and Early 1960s." *Government Publications Review* 16, no. 1 (1989): 15–29.

Tobey, Ronald C. *The American Ideology of National Science, 1919–1930*. Pittsburgh: University of Pittsburgh Press, 1971.

Traub, H. P., and H. J. Muller. "X-Ray Dosage in Relation to Germination of Pecan Nuts." *Botanical Gazette* 95, no. 4 (1934): 702–6.

Tribe, Derek. *Feeding and Greening the World: The Role of International Agricultural Research*. Wallingford, UK: CAB International, 1994.

"Tricking Dame Nature." *Muscatine Journal and News-Tribune*, 19 April 1939, 22. NewspaperArchive.

Turney, Jon. *Frankenstein's Footsteps: Science, Genetics, and Popular Culture*. New Haven, CT: Yale University Press, 1998.

Tuttle, William M., Jr., "The Birth of an Industry: The Synthetic Rubber 'Mess' in World War II." *Technology and Culture* 22, no. 1 (1981): 35–67.

"200 Attend Lecture on Atomic Gardening." *Youngstown Vindicator*, 14 March 1961, 17. Google News Archive.

United States Atomic Energy Commission. *Some Applications of Atomic Energy in Plant Science*. Washington, DC: GPO, 1952.

United States Congress, Joint Committee on Atomic Energy. *The Contribution of Atomic Energy to Agriculture: Hearings before the Subcommittee on Research and Development of the Joint Committee on Atomic Energy*. 83rd Cong., 2nd sess., 31 March and 1 April 1954.

———. *Progress Report on Atomic Energy Research: Hearings before the Subcommittee on Research and Development of the Joint Committee on Atomic Energy.* 84th Cong., 2nd sess., 4–8 June 1956.

University of Missouri Agricultural Experiment Station. "Solving Farm Problems by Research: One Year's Work, Agricultural Experiment Station (Report of the Director; July 1, 1925 to June 30, 1926)." *Missouri Agricultural Experiment Station Bulletin* 244 (1926).

———. "Solving Farm Problems by Research: One Year's Work, Agricultural Experiment Station (Report of the Director; July 1, 1926 to June 30, 1927)." *Missouri Agricultural Experiment Station Bulletin* 256 (1927).

University of Tennessee Agricultural Experiment Station. *Progress of Agricultural Research in Tennessee, 1961–1962, 74th and 75th Annual Reports of the Tennessee Agricultural Experiment Station.* Knoxville: University of Tennessee, 1962.

———. *Sixty-Second Annual Report, 1949.* Knoxville: University of Tennessee, 1949.

———. *Sixty-Seventh Annual Report, 1954, of the Tennessee Agricultural Experiment Station.* Knoxville: University of Tennessee, 1954.

———. *Sixty-Eighth Annual Report, 1955, of the Tennessee Agricultural Experiment Station.* Knoxville: University of Tennessee, 1955.

"Unusual Possibilities in Breeding." In *Yearbook of Agriculture 1936*, 183–206. Washington, DC: USDA, 1936.

"Use of a Chemical in Plant Breeding." *Science* 90, no. 2328 (1939): 8a.

UT-AEC Agricultural Research Laboratory. Oak Ridge, TN: UT-AEC, 1966.

"UT-AEC Research Program." *Tennessee Farm and Home Science*, April–June 1954, 3, 10.

van Dijck, José. *Imagenation: Popular Images of Genetics.* London: MacMillan, 1998.

van Harten, A. M. *Mutation Breeding: Theory and Practical Applications.* Cambridge: Cambridge University Press, 1998.

Vettel, Eric James. *Biotech: The Countercultural Origins of an Industry.* Philadelphia: University of Pennsylvania Press, 2006.

Waksman, Steve. "California Noise: Tinkering with Hardcore and Heavy Metal in Southern California." *Social Studies of Science* 34, no. 5 (2004): 675–702.

Waldron, Webb. "Turnips or Tulips: Which Are You Planting?" *American Magazine*, March 1933, 138–42.

Walker, J. Samuel. *Permissible Dose: A History of Radiation Protection in the Twentieth Century.* Berkeley: University of California Press, 2000.

Wang, Jessica. *American Scientists in an Age of Anxiety: Scientists, Anticommunism, and the Cold War.* Chapel Hill, NC: University of North Carolina Press, 1999.

Warden, Philip. "Chinese Seeds Sent Here for Atom Hotfoot." *Chicago Daily Tribune*, 2 July 1957, 4. ProQuest.

Ware, J. O. "Plant Breeding and the Cotton Industry." In *Yearbook of Agriculture 1936*, 657–744. Washington, DC: USDA, 1936.

"War Gardens Yield Less Costly Food." *Hartford Courant*, 29 December 1941, 16. ProQuest.

Warmke, H. E. "Experimental Polyploidy and Rubber Content in Taraxacum kok-saghyz." *Botanical Gazette* 106, no. 3 (1945): 316–24.

Warmke, H. E., and Harriet Davidson. "Polyploidy Investigations." In *Carnegie Institution of Washington Year Book No. 41*, 186–89. Washington, DC: Carnegie Institution of Washington, 1942.

———. "Polyploidy Investigations." In *Carnegie Institution of Washington Year Book No. 42*, 153–57. Washington, DC: Carnegie Institution of Washington, 1943.

Wassermann, F. "Obituary—Bernard Rudolf Nebel (1901–1963)." *Radiation Research* 29, no. 1 (1966): 150–51.

Way, Roger D. *A History of Pomology and Viticulture at Geneva*. Geneva, NY: New York State Agricultural Experiment Station, 1986.

Weart, Spencer R. *Nuclear Fear: A History of Images*. Cambridge, MA: Harvard University Press, 1988.

Wellensiek, S. J. "The Newest Fad, Colchicine, and Its Origins." *Chronica Botanica* 5 (1939): 15–17.

Westergaard, M. "Øjvind Winge. 1886–1964." *Biographical Memoirs of Fellows of the Royal Society* 10 (1964): 357–69.

Westwick, Peter J. *The National Labs: Science in an American System, 1947–1974*. Cambridge, MA: Harvard University Press, 2003.

White, Katharine S. "The Changing Rose, the Enduring Cabbage." *New Yorker*, 5 March 1960, 136.

Winkler, Allan M. *Life under a Cloud: American Anxiety about the Atom*. Urbana: University of Illinois Press, 1999.

Winston, Mark L. *Travels in the Genetically Modified Zone*. Cambridge, MA: Harvard University Press, 2002.

Wise, George. *Willis R. Whitney, General Electric and the Origins of U.S. Industrial Research*. New York: Columbia University Press, 1985.

Witkowski, Jan, ed. *Illuminating Life: Selected Papers from Cold Spring Harbor*. Vol. 1, *(1903–1969)*. Cold Spring Harbor: CSHL Press, 2000.

Witkowski, Jan A., and John R. Inglis, eds., *Davenport's Dream: 21st Century Reflections on Heredity and Eugenics*. Cold Spring Harbor: CSHL Press, 2008.

Wolfe, Audra J. "The Cold War Context of the Golden Jubilee, Or, Why We Think of Mendel as the Father of Genetics." *Journal of the History of Biology* 45, no. 3 (2012): 389–414.

"Wonder Plant-Drug Will Help You Grow Giant Vegetables." *Chillicothe Constitution Tribune*, 10 March 1949, 2. NewspaperArchive.

Woodburn, John H. *20th Century Bioscience: Professor O. J. Eigsti and the Seedless Watermelon*. Raleigh, NC: Pentland Press, 1999.

Woodley, R. G. "Open Irradiation Facilities for Botanical and Genetical Research." In *Engineering Compendium on Radiation Shielding*. Vol 3, *Shield Design and Engineering*, edited by R. G. Jaeger, E. P. Blizard, A. B. Chilton, M. Grotenhuis, A. Hönig, Th. A. Jaeger, and H. H. Eisenlohr, 89–94. Berlin/Heidelberg: Springer-Verlag, 1970.

Woolley, Edward Mott. "Where Electricity Gives Up Its Magic." *Popular Science Monthly*, November 1924, 45–47, 155–56.

Wright, John L., ed. *Possible Dreams: Enthusiasm for Technology in America*. Dearborn, MI: Henry Ford Museum & Greenfield Village, 1992.

Wright, Susan. "Legitimating Genetic Engineering." *Perspectives in Biology and Medicine* 44, no. 2 (2001): 235–47.

"X-Ray Produces Rare Gladiolus." *Elyria Chronicle Telegram*, 30 September 1935, 8. NewspaperArchive.

"X-Rays Are Found to Alter Heredity." *New York Times*, 27 September 1933. ProQuest.

"X-Rays Imperil Heredity but Speed Up Evolution." *Berkeley Daily Gazette*, 6 August 1927, 4. Google News Archive.

"X-Rays New Aid to Nurserymen." *Los Angeles Times*, 3 April 1928, 7. ProQuest.

"X-Ray Speeds Up Nature," *Spokane Spokesman-Review*, 19 January 1929, 5. Google News Archive.

"X-Ray Tests Expected to Effect New Plants, Better Animals." *Fairbanks Daily News Miner*, 6 September 1930, 6. NewspaperArchive.

"X-Ray to Breed Super Animals." *Frederick Post*, 19 November 1927, 6. NewspaperArchive.

"Your Results Will Be Shown at Home Show." *Cleveland Plain Dealer*, 13 November 1960, G1. Plain Dealer Archive.

Zachmann, Karin. "Atoms for Peace and Radiation for Safety—How to Build Trust in Irradiated Foods in Cold War Europe and Beyond." *History and Technology* 27, no. 1 (2011): 65–90.

———. "Peaceful Atoms in Agriculture and Food: How the Politics of the Cold War Shaped Agricultural Research Using Isotopes and Radiation in Post War Divided Germany." *Dynamis* 35, no. 2 (2015): 307–31.

———. "Risky Rays for an Improved Food Supply? National and Transnational Food Irradiation Research as a Cold War Recipe." Preprint 7. Munich: Deutsches Museum, 2013.

Zallen, Doris T. "The Rockefeller Foundation and Spectroscopy Research: The Programs at Chicago and Utrecht." *Journal of the History of Biology* 25, no. 1 (1992): 67–89.

Zeman, Scott C., and Michael A. Amundson. Introduction to *Atomic Culture: How We Learned to Stop Worrying and Love the Bomb*, edited by Scott C. Zeman and Michael A. Amundson, 1–9. Boulder: University of Colorado Press, 2004.

Zindel, Paul. *The Effect of Gamma Rays on Man-in-the-Moon Marigolds*. Reprint, New York: HarperTrophy, 2005.

Zorn, Eric. "Seedlesss Melons Took the Oblong Way to Success." *Chicago Tribune*, 31 August 1999. Accessed 24 February 2012. http://articles.chicagotribune.com/1999-08-31/news/9908310083 _1_seedless-chicago-state-oj.

Index

The letter *f* following a page number denotes a figure.